Calculus 1 Review in Bite-Size Pieces

By Kathryn Paulk
Copyright © 2023

Updated: 06/15/2025

Table of Contents

Introduction

This book will help students who are currently taking or planning to take a course in Calculus 1. For each topic, key equations are listed and followed by detailed examples.

This book has been formatted so that it is easy to read on both paperback and also on electronic devices with the Kindle app (laptop, iPad, Kindle E-reader, and iPhone).

<u>Calculus 1 Review</u>

Limits

<u>Simple Limits</u>

Definition of a Limit

$$\lim_{x \to a} f(x) = L$$

The limit of $f(x)$,
as x approaches a, equals L

As x approaches a,
from either the left or right,
The function approaches a value of L.

In other words …
   As x gets close to a from either side,
   the function gets close to L.
   The function may not be defined at $f(a)$.
   That's why we just get close to it.

To evaluate a limit, as x approaches a, first try to
simply substitute the value of a into the expression.

With rational expressions, care must be taken to avoid
making the denominator equal to zero. Try to
rearrange the expression to avoid dividing by zero.

Simple Limits for $\lim\limits_{x \to a} f(x)$ -- Ex. 1	
Substitute x with 5.	$\lim\limits_{x \to 5} 3x^2$ $= 3(5)^2 = 75$
Avoid zero denominator. Use algebra then substitute.	$\lim\limits_{x \to 3} \dfrac{x - 3}{x^2 - 9}$ $= \lim\limits_{x \to 3} \dfrac{(x - 3)}{(x - 3)(x + 3)}$ $= \lim\limits_{x \to 3} \dfrac{1}{(x + 3)} = \dfrac{1}{6}$
Avoid zero denominator. Use algebra then substitute.	$\lim\limits_{x \to 0} \dfrac{5x^3 + 4x^2}{x^2}$ $= \lim\limits_{x \to 0} \dfrac{5x + 4}{1}$ $= \lim\limits_{x \to 0} 5x + 4 = 4$
Substitute x with 0.	$\lim\limits_{x \to 0} \left(x^2 + \dfrac{\cos 3x}{100} \right)$ $= 0 + \dfrac{1}{100} = 0.01$
Use algebra to simplify then substitute x with 1.	$\lim\limits_{x \to 1} \dfrac{1 - \frac{1}{x}}{x - 1}$ $= \lim\limits_{x \to 1} \dfrac{\frac{1}{x}(x - 1)}{(x - 1)} = \dfrac{1}{1} = 1$

Simple Limits for $\lim\limits_{x \to a} f(x)$ -- Ex. 2	
Just substitute because denom. $\neq 0$	$\lim\limits_{x \to -1} \dfrac{1 - \frac{1}{x}}{x - 1}$ $= \dfrac{1 + 1}{-1 - 1} = \dfrac{2}{-2} = -1$
Use algebra to simplify then substitute x with 3.	$\lim\limits_{x \to 3} \dfrac{x^2 + 2x - 15}{x^2 - 9}$ $= \lim\limits_{x \to 3} \dfrac{(x - 3)(x + 5)}{(x - 3)(x + 3)}$ $= \lim\limits_{x \to 3} \dfrac{(x + 5)}{(x + 3)} = \dfrac{8}{6} = \dfrac{4}{3}$
Use algebra to simplify then substitute x with -2.	$\lim\limits_{x \to -2} \dfrac{x^2 - 3x - 10}{x + 2}$ $= \lim\limits_{x \to -2} \dfrac{(x + 2)(x - 5)}{(x + 2)}$ $= \lim\limits_{x \to -2} (x - 5) = -7$
Just substitute because denom. $\neq 0$	$\lim\limits_{x \to -2} \dfrac{x^2 - 3x - 10}{x - 2}$ $= \lim\limits_{x \to -2} \dfrac{4 + 6 - 10}{-4} = \dfrac{0}{-4} = 0$

One-Sided Limits

Definition of a Limit Using One-Sided Limits

$$\lim_{x \to a} f(x) = L$$

The limit of $f(x)$,

as x approaches a, equals L

IFF $\quad \lim_{x \to a^+} f(x) = L \quad$ <u>and</u> $\quad \lim_{x \to a^-} f(x) = L$

In other words ...

- When x approaches a from the right (a^+) the function approaches L.
- When x approaches a from the left (a^-) the function approaches L.

If both are true, then we can simply say ...

When x approaches a
the function approaches L.

To evaluate a one-sided limit, either graphically or algebraically, consider approaching a value slightly smaller or slightly larger than **a** from one side.

One Sided Limits — Ex. 1

One-sided limits for a piece-wise function $y = f(x)$

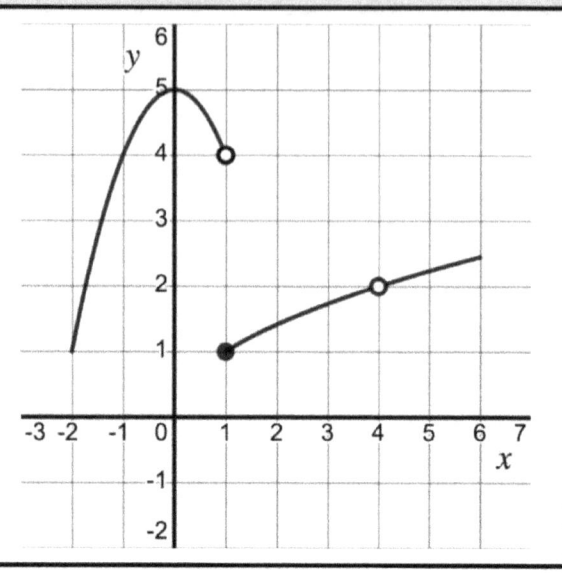

$$\lim_{x \to 0^-} f(x) = 5$$

$$\lim_{x \to 0^+} f(x) = 5$$

$$\lim_{x \to 0} f(x) = 5$$

$$\lim_{x \to 4^-} f(x) = 2$$

$$\lim_{x \to 4^+} f(x) = 2$$

$$\lim_{x \to 4} f(x) = 2$$

$$\lim_{x \to 1^-} f(x) = 4$$

$$\lim_{x \to 1^+} f(x) = 1$$

$$\lim_{x \to 1} f(x) = DNE$$

$$f(1) = 1$$

$$f(4) = DNE$$

$$f(0) = 5$$

Infinite Limits

Infinite Limit

$$\lim_{x \to a} f(x) = \infty$$

The limit of $f(x)$, as x approaches a,

from either side, becomes extremely large.

In other words ...

As x gets close to the value of a, from either side, the function becomes very, very large.

Note: ∞ is not a number. Since ∞ is not a number, there is actually no limit. The limit does NOT exist! The function is not limited to a particular number. It's just large (either positive or negative).

Infinite Limit (Positive Infinity)

$$\lim_{x \to a} f(x) = \infty$$

For every positive number M

there is a positive number δ Such that:

if $0 < |x - a| < \delta$ then $f(x) > M$

When x gets closer to a (from either side) the function will get larger than it was before.

Infinite Limit (Negative Infinity)

$$\lim_{x \to a} f(x) = -\infty$$

For every negative number N

there is a positive number δ Such that:

if $0 < |x - a| < \delta$ then $f(x) < N$

When x gets closer to a (from either side) the function will become more negative than it was before.

Vertical Asymptote (VA)

$x = a$ is a Vertical Asymptote (VA)

of $y = f(x)$

If at least one of the following is true:

$$\lim_{x \to a} f(x) = \infty$$

$$\lim_{x \to a^+} f(x) = \infty$$

$$\lim_{x \to a^-} f(x) = \infty$$

$$\lim_{x \to a} f(x) = -\infty$$

$$\lim_{x \to a^+} f(x) = -\infty$$

$$\lim_{x \to a^-} f(x) = -\infty$$

In other words ...

Infinite limits occur at vertical asymptotes.

Infinite Limit – Ex. 1

TIP: For one-sided limits, try substituting a number slightly larger or smaller than a. It will help you determine if the limit is positive or negative.

Try $x = 0.01$	$\displaystyle \lim_{x \to 0^+} \frac{1}{x^2} = \infty$
Try $x = -0.01$	$\displaystyle \lim_{x \to 0^-} \frac{1}{x^2} = \infty$
Try $x = 0.01$	$\displaystyle \lim_{x \to 0^+} \frac{1}{x} = \infty$
Try $x = -0.01$	$\displaystyle \lim_{x \to 0^-} \frac{1}{x} = -\infty$
Try $x = 3.01$	$\displaystyle \lim_{x \to 3^+} \frac{2x}{x-3} = \infty$
Try $x = 2.99$	$\displaystyle \lim_{x \to 3^-} \frac{2x}{x-3} = -\infty$
Try $x = 0.01$	$\displaystyle \lim_{x \to 0^+} \frac{1}{x^2} = \infty$
Tricky!	$\displaystyle \lim_{x \to 5^+} \frac{5-x}{x-5} = \lim_{x \to 5^+} \frac{-(x-5)}{x-5}$ $= \lim_{x \to 5^+} (-1) = -1$

Infinite Limit – Ex. 2

Infinite Limits occur at vertical asymptotes.

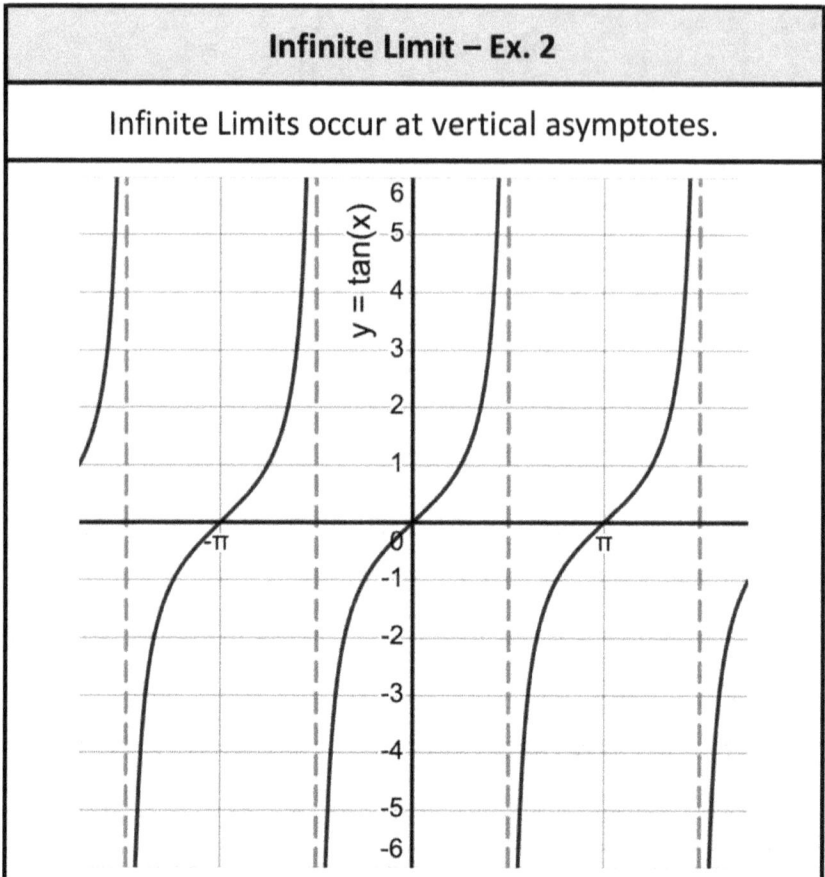

$\lim\limits_{x \to 0^-} \tan x = 0$	$\lim\limits_{x \to 0^+} \tan x = 0$
$\lim\limits_{x \to -\frac{\pi}{2}^-} \tan x = \infty$	$\lim\limits_{x \to -\frac{\pi}{2}^+} \tan x = -\infty$
$\lim\limits_{x \to \frac{\pi}{2}^-} \tan x = \infty$	$\lim\limits_{x \to \frac{\pi}{2}^+} \tan x = -\infty$

Limit Laws

Limit Laws
$\lim\limits_{x \to a} f(x) \quad = \quad f(a) \quad$ direct substitution
$\lim\limits_{x \to a} [\, f(x) \pm g(x)\,] \quad = \quad \lim\limits_{x \to a} f(x) \pm \lim\limits_{x \to a} g(x)$
$\lim\limits_{x \to a} [\, c \cdot f(x)\,] \quad = \quad c \cdot \lim\limits_{x \to a} f(x)$
$\lim\limits_{x \to a} [\, f(x) \cdot g(x)\,] \quad = \quad \lim\limits_{x \to a} f(x) \cdot \lim\limits_{x \to a} g(x)$
$\lim\limits_{x \to a} \left[\dfrac{f(x)}{g(x)} \right] \quad = \quad \dfrac{\lim\limits_{x \to a} f(x)}{\lim\limits_{x \to a} g(x)}$
$\lim\limits_{x \to a} [\, f(x)\,]^n \quad = \quad \left[\lim\limits_{x \to a} f(x) \right]^n$

Limits – Useful Equations

When evaluating limits, there are some simple limits that are useful to know. Later, you will learn techniques   (e.g. L'Hospital's Rule) that will help with evaluating limits.

For now, just memorize these three limits to help with evaluating a variety of limits.

$$\lim_{x \to 0} \frac{\sin x}{x} = 1$$

$$\lim_{x \to 0} \frac{1 - \cos x}{x} = 0$$

$$\lim_{x \to \infty} \frac{c}{x} = 0$$

The Squeeze Theorem

We know that if $a \leq b \leq c$ and also $a = c$

Then we can conclude that $a = b = c$.

This idea is applied to functions

in the Squeeze Theorem.

The Squeeze Theorem

If $\qquad f(x) \leq g(x) \leq h(x)$

And $\qquad \lim_{x \to a} f(x) = \lim_{x \to a} h(x) = L$

Then $\qquad \lim_{x \to a} g(x) = L$

Note: The Squeeze Theorem is helpful
with oscillating functions.

The Squeeze Theorem – Ex. 1

Evaluate: $\lim\limits_{x\to 0} [\, x^2 \cos x \,]$

-1	$\leq$	$\cos x$	$\leq$	1
$x^2\,(-1)$	$\leq$	$x^2 \cos x$	$\leq$	$x^2\,(1)$
$-x^2$	$\leq$	$x^2 \cos x$	$\leq$	x^2
$\lim\limits_{x\to 0} [-x^2]$	$\leq$	$\lim\limits_{x\to 0} [\, x^2 \cos x \,]$	$\leq$	$\lim\limits_{x\to 0} [x^2]$
0	$\leq$	$\lim\limits_{x\to 0} [\, x^2 \cos x \,]$	$\leq$	0

$\lim\limits_{x\to 0} [\, x^2 \cos x \,] = 0$

by The Squeeze Theorem

The Squeeze Theorem – Ex. 2

Evaluate: $\lim\limits_{x \to 0} \left[x^2 \sin\left(\frac{1}{x}\right) \right]$

$$-1 \leq \sin\frac{1}{x} \leq 1$$

$$x^2(-1) \leq x^2 \sin\left(\frac{1}{x}\right) \leq x^2(1)$$

$$-x^2 \leq x^2 \sin\left(\frac{1}{x}\right) \leq x^2$$

$$\lim_{x \to 0}[-x^2] \leq \lim_{x \to 0}\left[x^2 \sin\left(\frac{1}{x}\right)\right] \leq \lim_{x \to 0}[x^2]$$

$$0 \leq \lim_{x \to 0}\left[x^2 \sin\left(\frac{1}{x}\right)\right] \leq 0$$

$$\lim_{x \to 0}\left[x^2 \sin\left(\frac{1}{x}\right)\right] = 0$$

by The Squeeze Theorem

(Stewart, Calculus Early Transcendentals, p. 101)

Delta-Epsilon Definition of a Limit

Delta-Epsilon Definition of a Limit

$$\lim_{x \to a} f(x) = L$$

If   for every number $\varepsilon > 0$

there is a number $\delta > 0$

such that

If $0 < |x - a| < \delta$

then $|f(x) - L| < \varepsilon$

Notes:

- Delta (δ) is the error associated with x

- Epsilon (ε) is the error associated with y

Delta-Epsilon Definition of a Limit – Illustrated

$$\lim_{x \to a} f(x) = L \qquad \text{with } y = f(x) = x^2$$

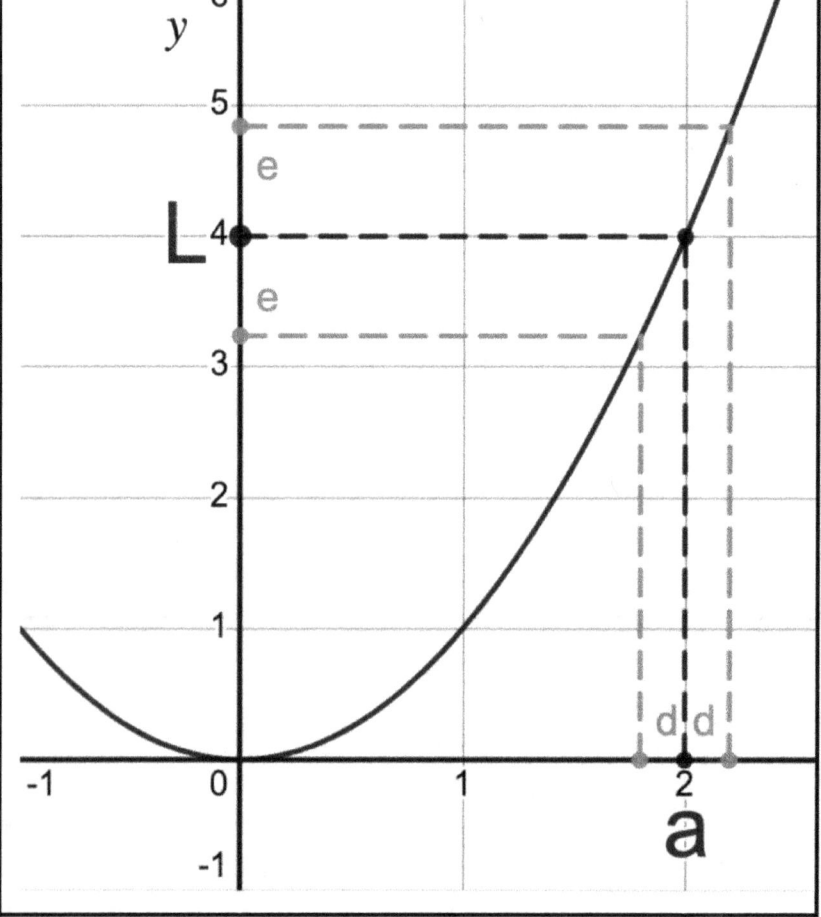

Left and Right-Handed Limits

Definition of a Left-Hand Limit

$$\lim_{x \to a^-} f(x) = L$$

If for every $\varepsilon > 0$ there is a number $\delta > 0$ such that

If $\quad a - \delta < x < a \quad$ then $\quad | f(x) - L | < \varepsilon$

Definition of a Right-Hand Limit

$$\lim_{x \to a^+} f(x) = L$$

If for every $\varepsilon > 0$ there is a number $\delta > 0$ such that

If $\quad a < x < a + \delta \quad$ then $\quad | f(x) - L | < \varepsilon$

The input and output errors can be more precisely defined if the input error is just one side of x.

Right-Hand Limit – Ex. 1

Prove: $\lim\limits_{x \to 4^+} \sqrt{x} = 2$ Here: $a = 4$, $L = 2$

$a < x < a + \delta$	$\lvert f(x) - L \rvert < \varepsilon$
$4 < x < 4 + \delta$	$\lvert \sqrt{x} - 2 \rvert < \varepsilon$
$0 < x - 4 < \delta$	$\sqrt{x} < \varepsilon + 2$
	$x < (\varepsilon + 2)^2$
	$x - 4 < (\varepsilon + 2)^2 - 4$

So, we should choose: $\delta = (\varepsilon + 2)^2 - 4$

$x < 4 + \delta$

$x < 4 + [(\varepsilon + 2)^2 - 4] = (\varepsilon + 2)^2$

$\sqrt{x} < \lvert \varepsilon + 2 \rvert$

$\lvert \sqrt{x} - 2 \rvert < \varepsilon$

$\lvert f(x) - L \rvert < \varepsilon \quad \rightarrow$

$$\lim\limits_{x \to a^+} f(x) = L$$

$$\lim\limits_{x \to 4^+} \sqrt{x} = 2$$

<u>Continuous Functions</u>

Continuous Function at a	
$$\lim_{x \to a} f(x) = f(a)$$	
$$\lim_{x \to a^+} f(x) = f(a)$$	Continuous from the right
$$\lim_{x \to a^-} f(x) = f(a)$$	Continuous from the left

In other words ...

- As x approaches a, from either side, the function approaches $f(a)$.
- There are NO gaps or jumps.

Notes

- Polynomial functions are continuous everywhere.
- Rational functions are continuous where defined.

Intermediate Value Theorem

Intermediate Value Theorem

If $f(x)$ is continuous on $[a, b]$

Where $f(a) < N < f(b)$

Then, there is a number c in (a, b)

Such that $f(c) = N$.

In other words ...

- Between two known values of a continuous function, the function will have another value between those two values.

Intermediate Value Theorem – Ex. 1

Given: $f(x) = x^3 - 6x + 2$

Show there is a root between 2 and 3.

Function is a polynomial so it is continuous.

A root is where: $f(c) = N = 0$

Check the boundaries:

$$f(2) = (2)^3 - 6(2) + 2 = -2$$
$$f(3) = (3)^3 - 6(3) + 2 = 11$$

$-2 < 0 < 11 \quad \rightarrow \quad$ A root is in interval $(2, 3)$

Note: $f(2)$ is negative and $f(3)$ is positive.

There is a sign change so there must be a root in the interval $(2, 3)$.

Intermediate Value Theorem – Ex. 2

Given: $f(x) = x^3 - 6x + 2$

Graphically show there is a root between 2 and 3.

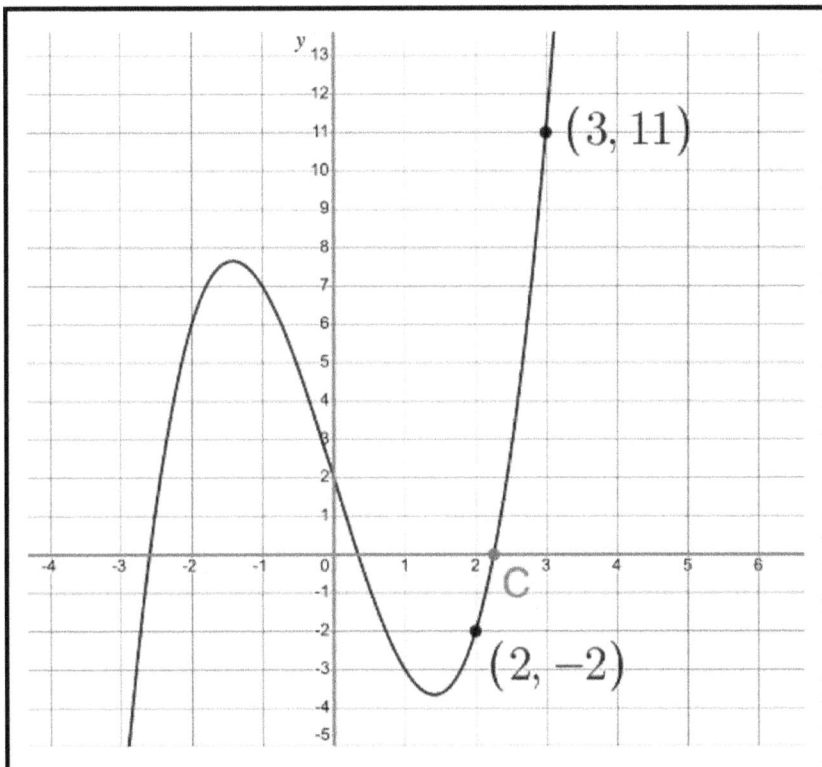

The graph shows 3 roots in 3 intervals:

$(-3, -2), \qquad (0, 1), \qquad and \; (2, 3)$

Limits at x --> Infinity

Limit at Infinity
$$\lim_{x \to \pm\infty} f(x) = L$$

In other words ...

- As x gets very large or very small,
  the function approaches a value of L.

Recall: At most, a function may have one horizontal asymptote (HA). If a function has a HA, then as x gets very large or very small, the function will approach the value of the HA. A horizontal asymptote describes a function's far-end behavior.

Limit at Infinity – Graphs

$f(x) = \dfrac{(x-1)}{(x^2+1)}$ $\displaystyle\lim_{x \to \pm\infty} f(x) = 0$	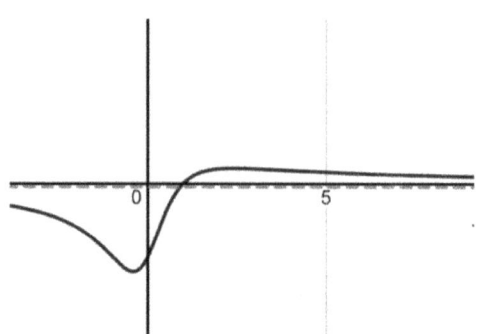
$f(x) = \dfrac{1}{x} + 2$ $\displaystyle\lim_{x \to \pm\infty} f(x) = 2$	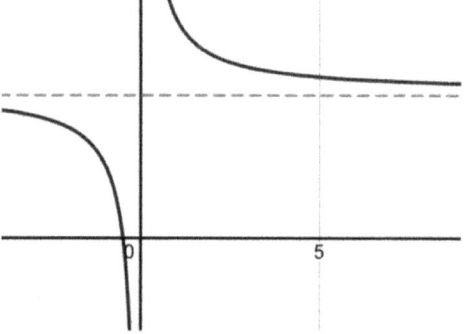
$f(x) = \sin\left(\dfrac{1}{x}\right)$ $\displaystyle\lim_{x \to \pm\infty} f(x) = 0$	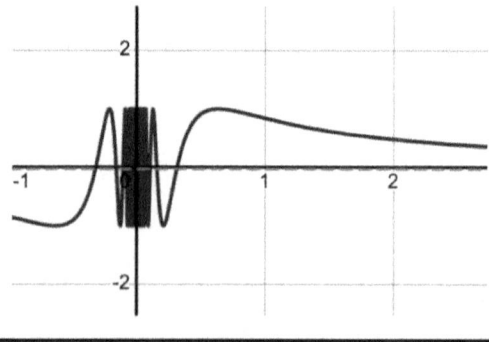

Limit at Infinity – Ex. 1

Graph the function: $y = \text{ArcTan}(x)$

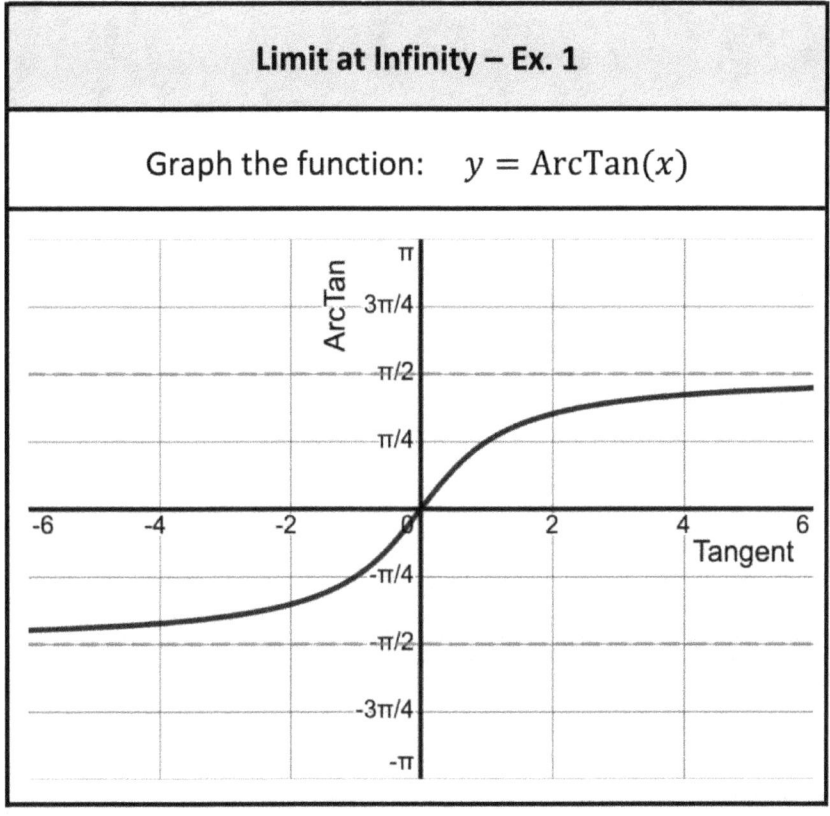

$$\lim_{x \to -\infty} \tan^{-1} x = -\frac{\pi}{2} \qquad \lim_{x \to \infty} \tan^{-1} x = \frac{\pi}{2}$$

Limit at Infinity – Ex. 2a

$$\lim_{x \to \infty} \frac{2x + 3}{\sqrt{4x^2 + 5}} \;=\; \frac{\infty}{\infty} \;=\; ???$$

$$= \lim_{x \to \infty} \frac{\left(\frac{1}{x}\right)}{\left(\frac{1}{x}\right)} \frac{2x + 3}{\sqrt{4x^2 + 5}} = \lim_{x \to \infty} \frac{\left|\frac{1}{x}\right|}{\sqrt{\frac{1}{x^2}}} \frac{2x + 3}{\sqrt{4x^2 + 5}}$$

$$= \lim_{x \to \infty} \frac{2 + \frac{3}{x}}{\sqrt{4 + \frac{5}{x^2}}} = \frac{2 + 0}{\sqrt{4 + 0}} = \frac{2}{2} = 1$$

$$\lim_{x \to -\infty} \frac{2x + 3}{\sqrt{4x^2 + 5}} \;=\; \frac{-\infty}{\infty} \;=\; ???$$

$$= \lim_{x \to -\infty} \frac{\left(\frac{1}{x}\right)}{\left(\frac{1}{x}\right)} \frac{2x + 3}{\sqrt{4x^2 + 5}} = \lim_{x \to \infty} \frac{\left|\frac{1}{x}\right|}{\sqrt{\frac{1}{x^2}}} \frac{2x + 3}{\sqrt{4x^2 + 5}}$$

$$= \lim_{x \to \infty} \frac{-2 - \frac{3}{x}}{\sqrt{4 + \frac{5}{x^2}}} = \frac{-2 - 0}{\sqrt{4 + 0}} = \frac{-2}{2} = -1$$

Limit at Infinity – Ex. 2b

Another way to look at limits with $x \to \pm\infty$ is to realize that we are looking for the far-end behavior, or the horizontal asymptote. Recall from pre-calculus, for rational functions, the HA is the ratio of the coefficients of the first terms.

For example: If $f(x) = \dfrac{ax^2 + bx + c}{dx^2 + gx + h}$

The degrees of the numerator and denominator are equal so the Horizontal Asymptote is at: $y = \dfrac{a}{d}$

$$\lim_{x \to \infty} \frac{2x + 3}{\sqrt{4x^2 + 5}} = \lim_{x \to \infty} \frac{2x}{\sqrt{4x^2}}$$

$$= \lim_{x \to \infty} \frac{2}{\sqrt{4}} \frac{x}{\sqrt{x^2}} = \lim_{x \to \infty} \frac{2}{2} \frac{x}{|x|} = 1$$

$$\lim_{x \to -\infty} \frac{2x + 3}{\sqrt{4x^2 + 5}} = \lim_{x \to -\infty} \frac{2x}{\sqrt{4x^2}}$$

$$= \lim_{x \to -\infty} \frac{2}{\sqrt{4}} \frac{x}{\sqrt{x^2}} = \lim_{x \to -\infty} \frac{2}{2} \frac{x}{|x|} = -1$$

Limit Review

Limit Review – Laws and Equations

$$\lim_{x \to a} f(x) = f(a) \quad \text{direct substitution}$$

$$\lim_{x \to a} [f(x) \pm g(x)] = \lim_{x \to a} f(x) \pm \lim_{x \to a} g(x)$$

$$\lim_{x \to a} [c \cdot f(x)] = c \cdot \lim_{x \to a} f(x)$$

$$\lim_{x \to a} [f(x) \cdot g(x)] = \lim_{x \to a} f(x) \cdot \lim_{x \to a} g(x)$$

$$\lim_{x \to a} \left[\frac{f(x)}{g(x)} \right] = \frac{\lim_{x \to a} f(x)}{\lim_{x \to a} g(x)}$$

$$\lim_{x \to a} [f(x)]^n = \left[\lim_{x \to a} f(x) \right]^n$$

$$\lim_{x \to 0} \frac{\sin x}{x} = 1$$

$$\lim_{x \to 0} \frac{1 - \cos x}{x} = 0$$

$$\lim_{x \to \infty} \frac{c}{x} = 0$$

Limit Review – Ex. 01
Evaluate: $\displaystyle\lim_{x \to \infty} \frac{x^2 + 5}{7 - 3x + 2x^3}$

Divide numerator and denominator by x^3	$\displaystyle\lim_{x \to \infty} \frac{x^2 + 5}{7 - 3x + 2x^3}$ $\displaystyle\lim_{x \to \infty} \frac{\dfrac{1}{x} + \dfrac{5}{x^3}}{\dfrac{7}{x^3} - \dfrac{3}{x^2} + 2}$
Substitute $x = \infty$ Then evaluate	$\dfrac{0 + 0}{0 - 0 + 2} = \dfrac{0}{2} = 0$

Limit Review – Ex. 02

Evaluate: $\quad \lim\limits_{x \to 5} f(x)$

Given: $\quad f(x) = \begin{cases} 3x + 2 & , \ x > 5 \\ 1 & , \ x = 5 \\ x^2 + 8 & , \ x < 5 \end{cases}$

Evaluate limit From the left.	$\lim\limits_{x \to 5^-} f(x) = 5^2 + 8$ $\lim\limits_{x \to 5^-} f(x) = 33$
Evaluate limit from the right.	$\lim\limits_{x \to 5^+} f(x) = 3(5) + 2$ $\lim\limits_{x \to 5^+} f(x) = 17$
Limit exists only if limits from left and right are equal.	$\lim\limits_{x \to 5} f(x) = DNE$ $DNE = $ Does Not Exist

Limit Review – Ex. 03	
Evaluate:	$\displaystyle\lim_{x \to 0} \frac{\tan(2x)}{x}$
Recall:	$\displaystyle\lim_{x \to 0} \frac{\sin x}{x} = 1$

Also, recall Trig. Double angles	$\sin(2u) = 2 \sin u \cdot \cos u$ $\cos(2u) = \cos^2 u - \sin^2 u$
Simplify. Use Trig. Substitutions	$\displaystyle\lim_{x \to 0} \frac{1}{x} \left[\frac{\sin(2x)}{\cos(2x)} \right]$ $\displaystyle\lim_{x \to 0} \frac{1}{x} \left[\frac{2 \sin x \cdot \cos x}{\cos^2 x - \sin^2 x} \right]$
Isolate the known limit.	$\displaystyle\lim_{x \to 0} \left(\frac{\sin x}{x} \right) \cdot \left[\frac{2 \cdot \cos x}{\cos^2 x - \sin^2 x} \right]$
Substitute $x = 0$ Then evaluate	$(1) \left[\frac{2 \cdot (1)}{1 - 0} \right] = \frac{2}{1} = 2$

Limit Review – Ex. 04	
Evaluate:	$\lim\limits_{x \to 0} \dfrac{1 - \cos x + 8x}{x}$
Recall:	$\lim\limits_{x \to 0} \dfrac{1 - \cos x}{x} = 0$

Rearrange	$\lim\limits_{x \to 0} \left[\dfrac{1 - \cos x}{x} + \dfrac{8x}{x} \right]$
Simplify	$\lim\limits_{x \to 0} \left[\dfrac{1 - \cos x}{x} + 8 \right]$
Evaluate	$\lim\limits_{x \to 0} \left(\dfrac{1 - \cos x}{x} \right) + \lim\limits_{x \to 0} \left(8 \right)$ $0 + 8 = 8$

Limit Review – Ex. 05

Evaluate: $\lim\limits_{x \to 0} \dfrac{|x|}{x}$

Evaluate from the left. Note: x is negative	$\lim\limits_{x \to 0} \dfrac{|x|}{x}$ $\dfrac{positive}{negative} = -1$
Evaluate from the right Note: x is positive	$\lim\limits_{x \to 0} \dfrac{|x|}{x} = \dfrac{positive}{positive} = 1$
A limit exists only if the limit from the left & right are equal.	$\lim\limits_{x \to 0} \dfrac{|x|}{x} = DNE$ $DNE =$ Does Not Exist

Limit Review – Ex. 06

Evaluate: $\displaystyle\lim_{x \to 3^-} \frac{x + 1}{x - 3}$

Evaluate this limit from the left for $x \approx 2.9$ 2.9 is a little less than 3.	$\dfrac{3 + 1}{2.9 - 3}$ $\dfrac{4}{negative\ zero} = -\infty$

Note: When looking at the original limit, it is easy to see it is in the form: $\frac{c}{0} = \pm\infty$.

For this one-sided limit, the denominator is negative as it approaches zero.

Limit Review – Ex. 07

Evaluate: $\lim\limits_{x \to 3} f(g(x))$

Given: $f(x)$ and $g(x)$ are both continuous functions with the following table of values.

x	1	2	3	4
$f(x)$	3	6	9	12
$g(x)$	0	1	2	3

Work from the inside to the outside.	$\lim\limits_{x \to 3} f(g(x))$ $f\left(\lim\limits_{x \to 3} g(x) \right)$ $f(g(3))$ $f(2) = 6$

Limit Review – Ex. 08

Given the graph of $f(x)$, evaluate the limits.

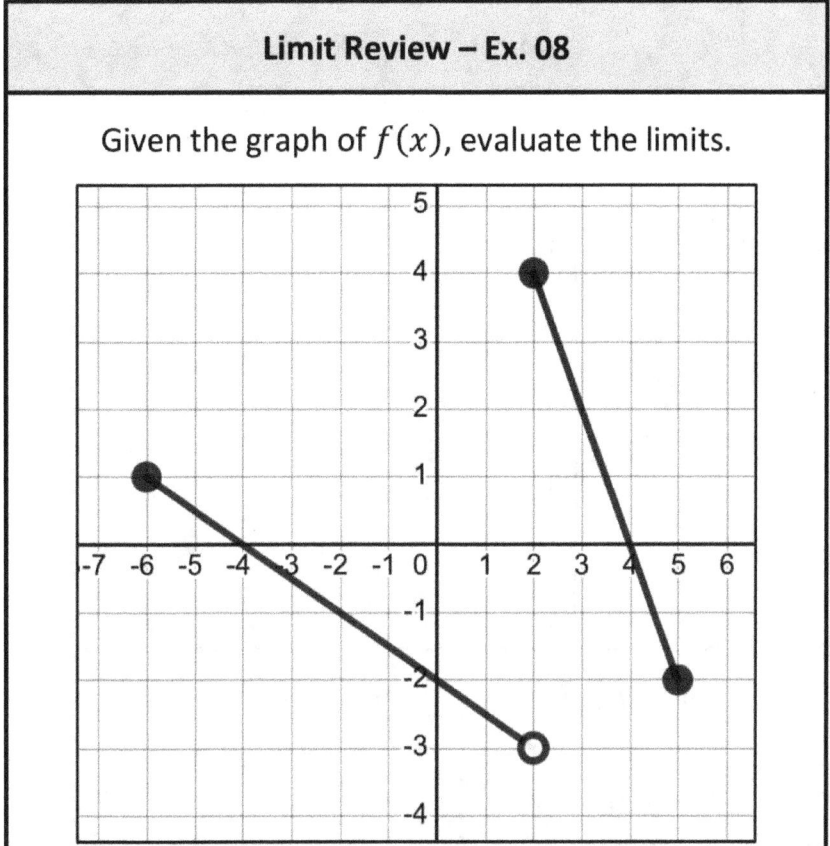

$\lim\limits_{x \to 2^-} f(x)$	$\lim\limits_{x \to 2^-} f(x) = -3$
$\lim\limits_{x \to 2^+} f(x)$	$\lim\limits_{x \to 2^+} f(x) = 4$
$\lim\limits_{x \to 2} f(x)$	$\lim\limits_{x \to 2} f(x) = DNE$

Limit Review – Ex. 09

Given:
the graph
of $f(x)$,

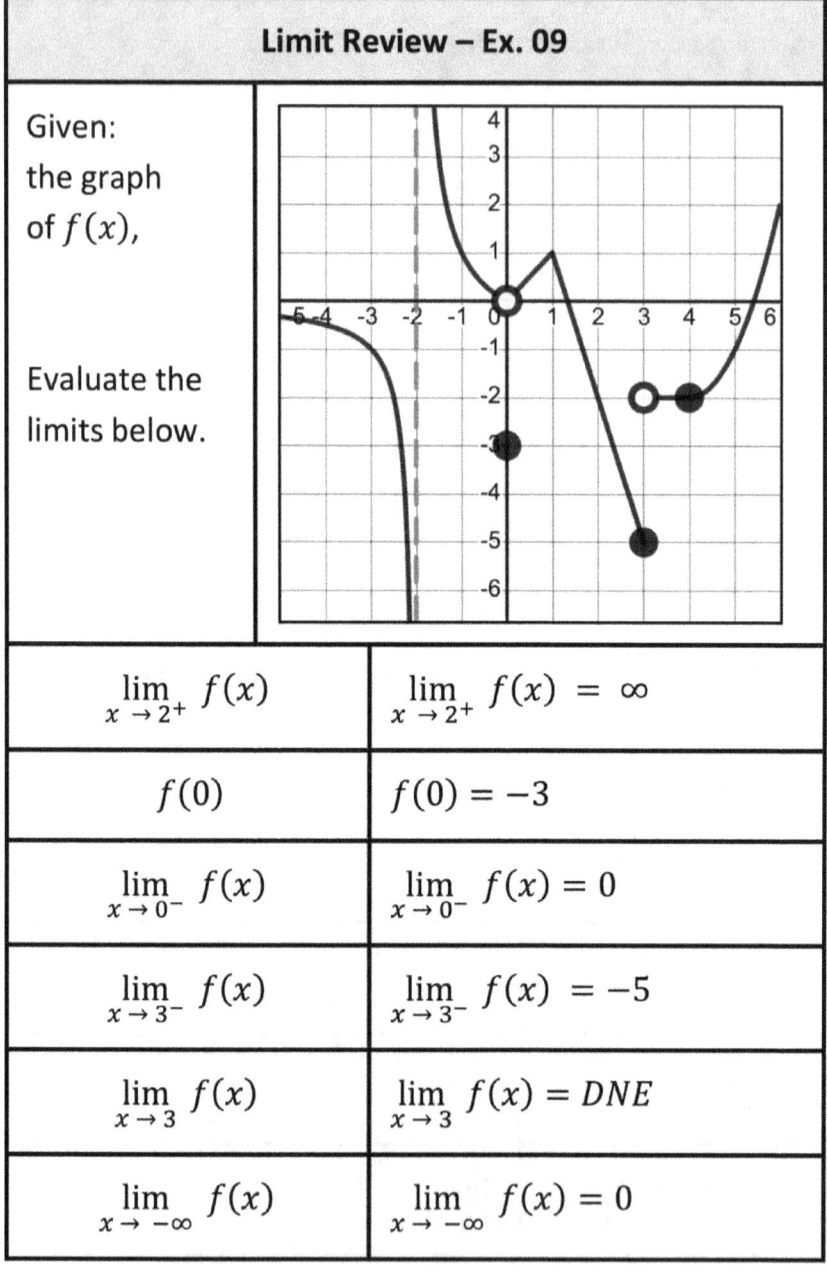

Evaluate the
limits below.

$\lim\limits_{x \to 2^+} f(x)$	$\lim\limits_{x \to 2^+} f(x) = \infty$
$f(0)$	$f(0) = -3$
$\lim\limits_{x \to 0^-} f(x)$	$\lim\limits_{x \to 0^-} f(x) = 0$
$\lim\limits_{x \to 3^-} f(x)$	$\lim\limits_{x \to 3^-} f(x) = -5$
$\lim\limits_{x \to 3} f(x)$	$\lim\limits_{x \to 3} f(x) = DNE$
$\lim\limits_{x \to -\infty} f(x)$	$\lim\limits_{x \to -\infty} f(x) = 0$

Limit Review – Ex. 10

Given: $\lim\limits_{x \to a} f(x) = 20$ and $\lim\limits_{x \to a} g(x) = -5$

Evaluate: The following limits.

$\lim\limits_{x \to a} [g(x) - f(x)]$	$\lim\limits_{x \to a} g(x) - \lim\limits_{x \to a} f(x)$ $-5 - 20 = -25$
$\lim\limits_{x \to a} [2f(x) + g(x)]$	$2 \lim\limits_{x \to a} f(x) + \lim\limits_{x \to a} g(x)$ $2(20) + (-5)$ $40 - 5 = 35$
$\lim\limits_{x \to a} [g(x) \cdot f(x)]$	$\left[\lim\limits_{x \to a} g(x) \right] \cdot \left[\lim\limits_{x \to a} f(x) \right]$ $(-5) \cdot (20) = -100$

Limit Review – Ex. 11

Evaluate: $\quad \lim\limits_{x \to -2} \dfrac{x^2 + 5x + 6}{x^2 - 5x - 10}$

Factor the numerator and denominator.	$\lim\limits_{x \to -2} \dfrac{(x+2)(x+3)}{(x+2)(x-5)}$
Cancel terms	$\lim\limits_{x \to -2} \dfrac{(x+3)}{(x-5)}$
Use substitution to evaluate.	$\dfrac{(-2+3)}{(-2-5)} = -\dfrac{1}{7}$

Limit Review – Ex. 12
Evaluate: $\displaystyle\lim_{x \to 1} \dfrac{\sqrt{3x + 2} - \sqrt{5}}{x - 1}$

Eliminate radicals in the numerator.	$\displaystyle\lim_{x \to 1} \dfrac{\sqrt{3x + 2} - \sqrt{5}}{x - 1} \cdot \left[\dfrac{\sqrt{3x + 2} + \sqrt{5}}{\sqrt{3x + 2} + \sqrt{5}} \right]$ $\displaystyle\lim_{x \to 1} \dfrac{(3x + 2) - (5)}{(x - 1) \cdot (\sqrt{3x + 2} + \sqrt{5})}$ $\displaystyle\lim_{x \to 1} \dfrac{3x - 3}{(x - 1) \cdot (\sqrt{3x + 2} + \sqrt{5})}$
Simplify	$\displaystyle\lim_{x \to 1} \dfrac{3(x - 1)}{(x - 1) \cdot (\sqrt{3x + 2} + \sqrt{5})}$ $\displaystyle\lim_{x \to 1} \dfrac{3}{(\sqrt{3x + 2} + \sqrt{5})}$
Use substitution to evaluate.	$\dfrac{3}{\left(\sqrt{3(1) + 2} + \sqrt{5}\right)}$ $\dfrac{3}{(\sqrt{5} + \sqrt{5})} = \dfrac{3}{2\sqrt{5}}$

Limit Review – Ex. 13

Evaluate: $\displaystyle \lim_{x \to 0} \frac{\dfrac{1}{x+5} - \dfrac{1}{5}}{x}$

Write numerator as a single fraction.	$\displaystyle \lim_{x \to 0} \frac{\dfrac{1}{x+5}\left(\dfrac{5}{5}\right) - \dfrac{1}{5}\left(\dfrac{x+5}{x+5}\right)}{x}$
	$\displaystyle \lim_{x \to 0} \frac{\left(\dfrac{5-(x+5)}{5(x+5)}\right)}{x}$
Then simplify	$\displaystyle \lim_{x \to 0} \frac{\left(\dfrac{-x}{5(x+5)}\right)}{x}$
	$\displaystyle \lim_{x \to 0} \frac{-x}{5x(x+5)}$
	$\displaystyle \lim_{x \to 0} \frac{-1}{5(x+5)}$
Evaluate	$\displaystyle \frac{-1}{5(0+5)} = \frac{-1}{5(5)} = -\frac{1}{25}$

Limit Review – Ex. 14
Given: $5x - 4 \leq f(x) \leq x^2 + 2$ Find: $\displaystyle\lim_{x \to 2} f(x)$ and Justify your answer.

$5x - 4 \quad \leq \quad f(x) \quad \leq \quad x^2 + 2$
$\displaystyle\lim_{x \to 2} (5x - 4) \quad \leq \quad \lim_{x \to 2} f(x) \quad \leq \quad \lim_{x \to 2} (x^2 + 2)$
$(5(2) - 4) \quad \leq \quad \displaystyle\lim_{x \to 2} f(x) \quad \leq \quad (2^2 + 2)$ $(10 - 4) \quad \leq \quad \displaystyle\lim_{x \to 2} f(x) \quad \leq \quad (4 + 2)$ $6 \quad \leq \quad \displaystyle\lim_{x \to 2} f(x) \quad \leq \quad 6$
Conclusion: $\quad \displaystyle\lim_{x \to 2} f(x) = 6 \qquad$ By the Squeeze Theorem

Limit Review – Ex. 15

Given: $f(x) = \begin{cases} x^2 + k & , \ x \le 0 \\ 5 - \sin x & , \ x > 0 \end{cases}$

Find: k that will make $f(x)$ continuous at $x = 0$.
Use definition of continuity to justify your answer.

Find the k that makes each part equal at the boundary.	$x^2 + k = 5 - \sin x$ $\lim\limits_{x \to 0^-} (x^2 + k)$ $\lim\limits_{x \to 0^+} (5 - \sin x)$ $(0 + k) = (5 - 0) \quad \to \quad k = 5$
Recall the 3 conditions of continuity at $x = a$.	$\lim\limits_{x \to a^-} f(x)$ *exists* $\lim\limits_{x \to a^+} f(x)$ *exists* $\lim\limits_{x \to a} f(x) = f(a)$
Three conditions of continuity at $x = a$.	$\lim\limits_{x \to 0^-} f(x) = 0^2 + 5 = 5$ $\lim\limits_{x \to 0^+} f(x) = 5 - \sin 0 = 5$ $f(0) = 0^2 + 5 = 5$ So, $f(x)$ is continuous at $x = 0$

Limit Review – Ex. 16a

Given: $f(x) = x^2 - 2x - 1$ on $[-2, 10]$

Does the Intermediate Value Theorem (IVT)
Guarantee there is a value $x = c$
such that $f(c) = 23$? Justify your answer.

Recall, the Intermediate Value Theorem	If $f(x)$ is continuous on $[a, b]$ Where $f(a) < N < f(b)$ Then there is a number, c, in (a, b) where $f(c) = N$
Continuous?	$f(x)$ is a polynomial so it is cont.
Evaluate $f(x)$ at the boundaries.	$f(-2) = (-2)^2 - 2(-2) - 1 = 7$ $f(10) = (10)^2 - 2(10) - 1 = 79$ $7 < N < 79$ True when $N = 23$ Yes, it's guaranteed!

Continued ...

Limit Review – Ex. 16b

Given: $f(x) = x^2 - 2x - 1$ on $[-2, 10]$

Does the Intermediate Value Theorem (IVT)

Guarantee there is a value $x = c$

such that $f(c) = 23$?   Justify your answer.

Previously found	$N = f(c) = 23$ is guaranteed By the IVT
Find c where $f(c) = 23$	$c^2 - 2c - 1 = 23$ $c^2 - 2c - 24 = 0$ $(c - 6)(c + 4) = 0$ $c = 6, -4$.
Consider the interval $[-2, 10]$	-4 is not in the interval. Therefore, $c = 6$
Conclusion	$f(6) = 23$

Tangent Lines

Tangent Lines

Tangent lines are straight lines that

- Touch a curved line at one point.

- Touch a straight line at all points.

A good way to think of a tangent is to visualize a straw, leaning against a function, and touching it at exactly one point.

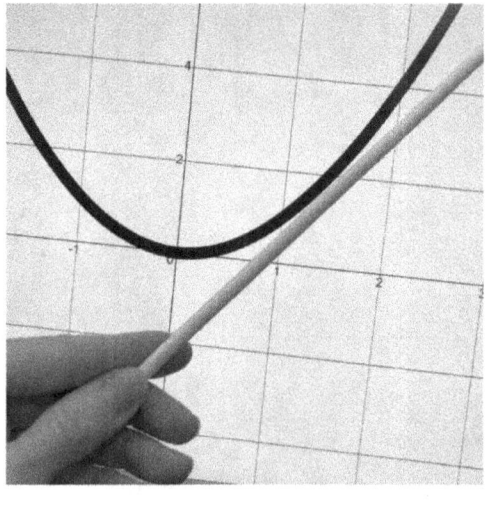

Tangent Lines – Ex. 1

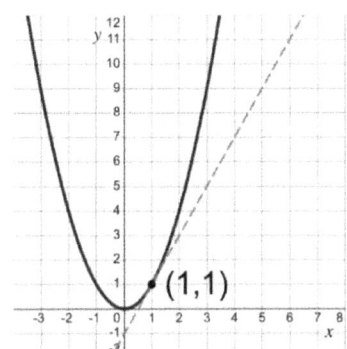

On: $y = x^2$

Tangent line has a

positive slope

at $x = 1$

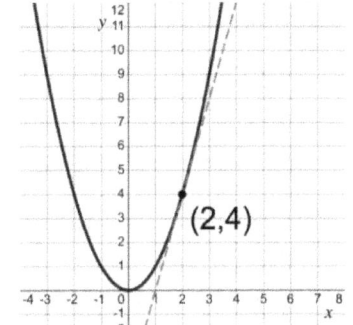

On: $y = x^2$

Tangent line has a

positive slope

at $x = 2$

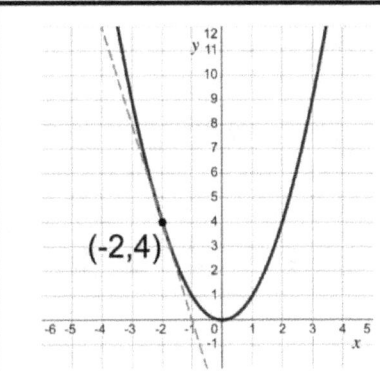

On: $y = x^2$

Tangent line has a

negative slope

at $x = -2$

Tangent Lines – Ex. 2

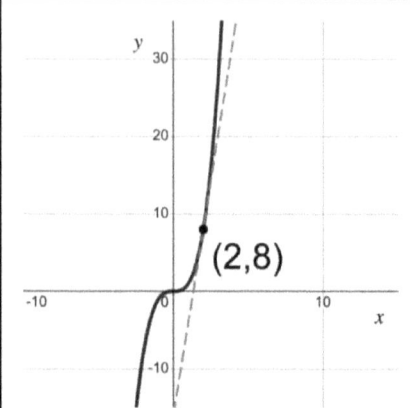

On: $y = x^3$

Tangent line has a

positive slope

at $x = 2$

(2,8)

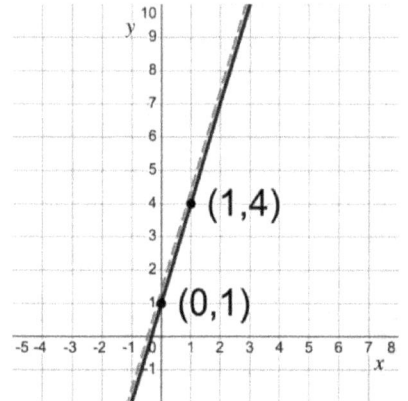

On: $y = 3x + 1$

Tangent line has a

positive slope

for all x

(1,4)

(0,1)

$y = mx + b$
$y = 3x + 1$

$m = slope$
$b = y \ intercept$

$m = slope$

$m = \dfrac{y_2 - y_1}{x_2 - x_1}$

$m = \dfrac{4 - 1}{1 - 0} = 3$

Approximate Slope of Tangent Line

Approximate Slope of Tangent Line

Using 2 Points on the curve, near $x = a$

$$Slope \approx \frac{f(a+h) - f(a)}{h}$$

In other words ...

The slope of a straight line can be calculated with:

$$m = \frac{\Delta y}{\Delta x} = \frac{y_2 - y_1}{x_2 - x_1}.$$

Instead of using two points on the tangent line, we are using two points on the curve, near $x = a$.

If h is small, then the two points are close.

Note: Two points on the curve, near $x = a$ are:

$$(a, f(a)) \quad \text{and} \quad (a + h, \ f(a + h))$$

Slope of Tangent Line
Approximated With 2 Points

Tangent line touches
a curve at ONE point .

$$(a, f(a))$$

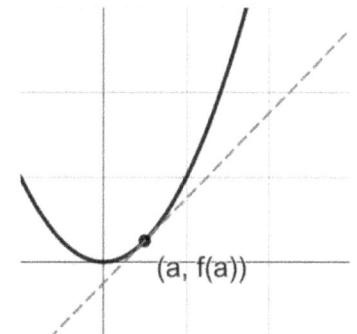

Use TWO points on the
curve to estimate slope
of the tangent line.

$$\text{Slope} = \frac{y_2 - y_1}{x_2 - x_1}$$

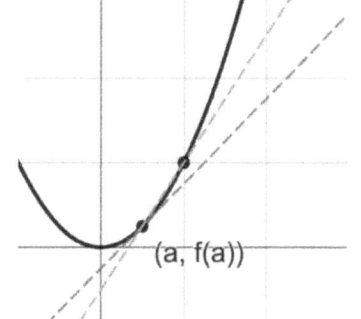

If the TWO points
are closer,
the estimate
is better.

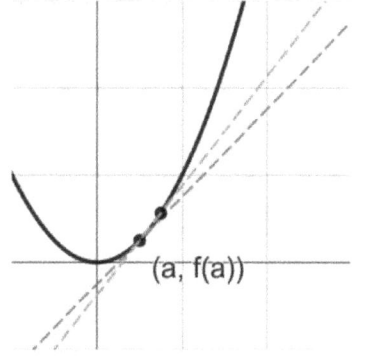

Linear Equations

Linear Equation Review
Equations of tangent lines are linear. So, a review of linear equations and some examples may be helpful.

Form	Linear Equation
Slope Intercept	$y = mx + b$
Point-Slope	$(y - y_1) = m(x - x_1)$
General or Standard	$Ax + By = C$

m	Slope	$m = \dfrac{\Delta y}{\Delta x} = \dfrac{(y_2 - y_1)}{(x_2 - x_1)}$
b	y-intercept	$b = y - mx$ $b = y_1 - mx_1$

Linear Equation Review -- Ex. 1

Given the slope of a line $(m = 3)$ and one point

$(4 , 17)$ on the line, find the equation of the line.

Use slope equation	$m = \dfrac{(y - y_1)}{(x - x_1)}$
Plug in given information	$3 = \dfrac{(y - 17)}{(x - 4)}$
Rearrange	$3(x - 4) = (y - 17)$ $(y - 17) = 3(x - 4)$ $y = 3x + 5$

Notes:

- Here, the given point is $(x_1, y_1) = (4, 17)$
- Point (x, y) is any point on the line.

Linear Equation Review -- Ex. 2

Given the slope of a line $(m = 3)$ and one point
$(4, 17)$ on the line, find the equation of the line.
Use a different solution than was used for Ex. 1

Use the equation	$y = mx + b$
Plug in given information	$y = 3x + b$
Solve for b Use given point. $(x, y) = (4,17)$	$b = y - 3x$ $b = 17 - 3(4)$ $b = 5$
Write the equation	$y = 3x + 5$

Linear Equation Review -- Ex. 3

Given two points: $(1, 8)$ and $(10, 35)$

Find the slope of line passing through them.
Then find the equation of the line.

Find the slope.	$m = \dfrac{y_2 - y_1}{x_2 - x_1}$ $m = \dfrac{35 - 8}{10 - 1} = \dfrac{27}{9} = 3$
Use point $(1, 8)$ to write the equation.	$(y - y_1) = m(x - x_1)$ $(y - 8) = 3(x - 1)$
Extra: Rearrange to get form: $y = mx + b$	$y - 8 = 3x - 3$ $y = 3x + 5$

Derivative at x = a

Limit Definition of a Derivative at $x = a$

The derivative at $x = a$, is the slope of the tangent line to the curve at $x = a$. Also known as the instantaneous rate of change.

$$f'(a) \;=\; \lim_{h \to 0} \frac{f(a+h) - f(a)}{h}$$

In other words …The derivative of a function, **where** $x = a$, is the slope between two very close points. In fact, the distance between them approaches zero!

$$f'(x) \;\approx\; \frac{y_2 - y_1}{x_2 - x_1} = \frac{\Delta y}{\Delta x} = \frac{dy}{dx}$$

$$f'(x) \;\approx\; \frac{f(x_2) - f(x_1)}{x_2 - x_1} \qquad \boxed{\begin{array}{l} x_1 = a \\ x_2 = a + h \end{array}}$$

$$f'(a) \;\approx\; \frac{f(a+h) - f(a)}{(a+h) - (a)} = \frac{f(a+h) - f(a)}{h}$$

$$f'(a) \;=\; \lim_{h \to 0} \frac{f(a+h) - f(a)}{h}$$

Limit Definition of a Derivative at $x = a$ -- Ex. 1

Use the limit definition of a derivative at $x = 2$

to find $f'(2)$ for $f(x) = x^2 - 1$

$$f'(a) = \lim_{h \to 0} \frac{f(a+h) - f(a)}{h}$$

$$f'(2) = \lim_{h \to 0} \frac{\left[(2+h)^2 - 1\right] - \left[2^2 - 1\right]}{h}$$

$$f'(2) = \lim_{h \to 0} \frac{\left[4 + 4h + h^2 - 1\right] - \left[3\right]}{h}$$

$$f'(2) = \lim_{h \to 0} \frac{4h + h^2}{h}$$

$$f'(2) = \lim_{h \to 0} 4 + h = 4 + 0$$

$$f'(2) = 4 \quad \text{Slope of tangent line at } x = 2$$

Derivative at any x

Limit Definition of a Derivative at $x = any\ x$

The derivative at any x, is the slope of the tangent line to the curve at any x. It is a function of x.

$$f'(x) = \lim_{h \to 0} \frac{f(x+h) - f(x)}{h}$$

In other words ...The derivative of a function at any x is the slope between x and another, very close point. In fact, the distance between them approaches zero!

$$f'(x) \approx \frac{y_2 - y_1}{x_2 - x_1} = \frac{\Delta y}{\Delta x} = \frac{dy}{dx}$$

$$f'(x) \approx \frac{f(x_2) - f(x_1)}{x_2 - x_1} \qquad \boxed{\begin{array}{l} x_1 = x \\ x_2 = x + h \end{array}}$$

$$f'(x) \approx \frac{f(x+h) - f(x)}{(x+h) - (x)} = \frac{f(x+h) - f(x)}{h}$$

$$f'(x) = \lim_{h \to 0} \frac{f(x+h) - f(x)}{h}$$

Limit Definition of a Derivative at $x = x$ — Ex. 1

Use the limit def. of a derivative at any x to find $f'(x)$ for $f(x) = x^2 - 1$

$$f'(x) = \lim_{h \to 0} \frac{f(x+h) - f(x)}{h}$$

$$f'(x) = \lim_{h \to 0} \frac{\left[(x+h)^2 - 1\right] - \left[x^2 - 1\right]}{h}$$

$$f'(x) = \lim_{h \to 0} \frac{\left[x^2 + 2xh + h^2 - 1\right] - \left[x^2 - 1\right]}{h}$$

$$f'(x) = \lim_{h \to 0} \frac{x^2 + 2xh + h^2 - 1 - x^2 + 1}{h}$$

$$f'(x) = \lim_{h \to 0} \frac{2xh + h^2}{h}$$

$$f'(x) = \lim_{h \to 0} 2x + h = 2x + 0$$

$$f'(x) = 2x \quad \text{Slope of tangent line at \underline{any x}}$$

$$f'(2) = 2(2) = 4 \qquad \text{Slope at } x = 2$$

Limit Definition of a Derivative at $x = x$ **– Ex. 2**

Find equation of the tangent line at $x = 2$
To the curve: $f(x) = x^2 - 1$

Find slope at $x = 2$	$m = 4$	Previously Found
Find coordinates when $x = 2$	$y = x^2 - 1$ $y = (2)^2 - 1$ $y = 3$	Given $\rightarrow \quad (x, y) = (2, 3)$
Find the y-intercept	$y = mx + b$ $b = y - mx$ $b = 3 - (4)2$ $b = -5$	
Equation of tangent line at $x = 2$	$y = mx + b$ $y = 4x - 5$	

Limit Definition of a Derivative at $x = x$ **–** **Ex. 3**

Graph the function: $f(x) = x^2 - 1$

and tangent line, previously found: $y = 4x - 5$

at the point $(x, y) = (2, 3)$

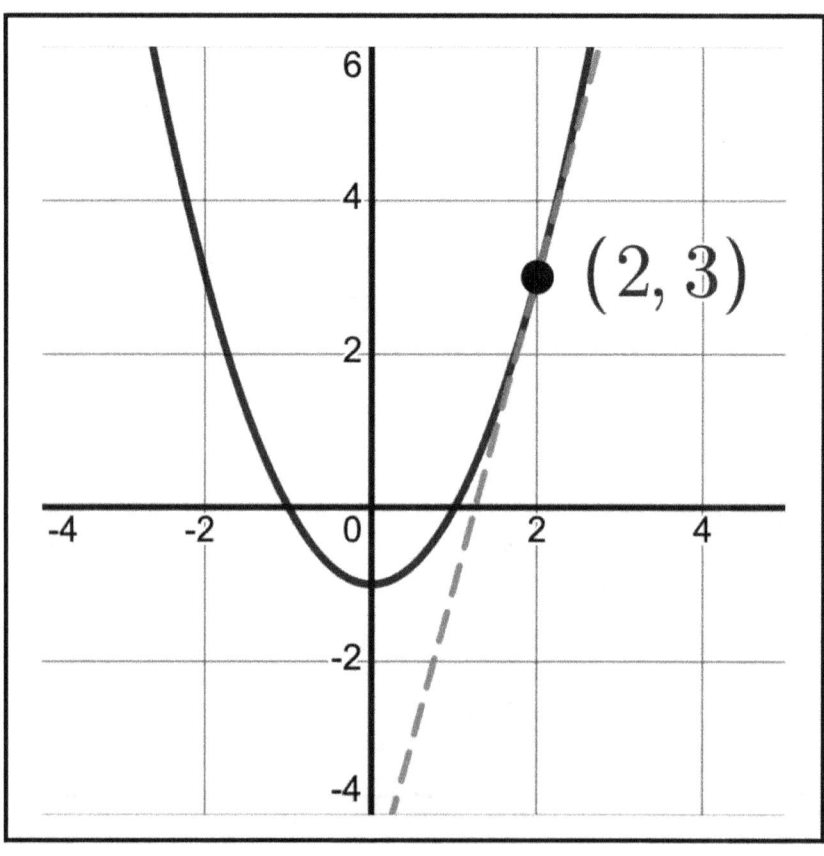

Derivatives

Basic Functions

Differentiation -- Conventions
$u = u(x)$ and $v = v(x)$ and $c = \text{constant}$

u'	$= \dfrac{d}{dx}[u] = \dfrac{du}{dx} = $ derivative of u
v'	$= \dfrac{d}{dx}[v] = \dfrac{dv}{dx} = $ derivative of v

Differentiation – Basic Functions	
Constant Multiplier	$\dfrac{d}{dx}[cu] \quad = cu'$
Sum and Difference	$\dfrac{d}{dx}[u \pm v] \quad = u' \pm v'$
Constant	$\dfrac{d}{dx}[c] \quad = 0$

Power Rule

*** **Differentiation – Power Rule** ***

$$\frac{d}{dx}(x^n) = nx^{n-1}$$

Power Rule – Example 0

Given: $f(x) = x^5$ Find the derivative of $f(x)$

$$y = f(x)$$

$$\frac{d}{dx}[y] = \frac{d}{dx}[f(x)]$$

$$\frac{d}{dx}[y] = \frac{d}{dx}[x^5]$$

$$\frac{dy}{dx} = \frac{d}{dx}[x^5]$$

$$y' = \frac{d}{dx}[x^5]$$

$$y' = 5x^4$$

$$f'(x) = 5x^4$$

Differentiation -- Power Rule -- Examples	
Given: $f(x)$	Use the Power Rule to find: $f'(x)$
Power Rule	$\frac{d}{dx}[x^n] = n \cdot x^{n-1}$

$f(x) = x^3$	$f'(x) = 3x^2$
$f(x) = \frac{1}{x^5}$	$f(x) = x^{-5}$ $f'(x) = -5x^{-6}$
$f(x) = \sqrt[7]{x^8}$	$f(x) = x^{\frac{7}{8}}$ $f'(x) = \frac{7}{8}x^{\left(\frac{7}{8}-\frac{8}{8}\right)}$ $f'(x) = \frac{7}{8}x^{\left(-\frac{1}{8}\right)}$
$f(x) = \frac{3x^5+6}{7}$	$f(x) = \left(\frac{3}{7}\right)x^5 + \left(\frac{6}{7}\right)$ $f'(x) = \left(\frac{3}{7}\right)5\,x^4 + 0$ $f'(x) = \left(\frac{15}{7}\right)x^4$

Chain Rule

Chain Rule for $F(x) = f(g(x))$

$$F'(x) = f'(g(x)) \cdot g'(x)$$

In Leibniz notation ...

$$\frac{dy}{dx} = \frac{dy}{du} \cdot \frac{du}{dx}$$

In other words ...

If an equation is in the form of a composite function, then we can use the chain rule.

Chain Rule – Simple Example

Given: $f(x) = (2x^3)^5$ Find: $f'(x)$

$$f(x) = (2x^3)^5$$

$$f'(x) = 5(2x^3)^4 \cdot 6x^2$$

$$f'(x) = 30(2x^3)^4 \cdot x^2$$

$$f'(x) = 30(16x^{12}) \cdot x^2$$

$$f'(x) = 480\,x^{14}$$

Chain Rule -- Examples	
Given: $f(x)$	Find: $f'(x)$

$f = (x^5 + 7x)^3$	$f' = 3(x^5 + 7x)^2(5x^4 + 7)$
$f = \dfrac{7}{\sqrt{5x^2 + 8}}$	$f = 7(5x^2 + 8)^{-\frac{1}{2}}$ $f' = -\dfrac{7}{2}(5x^2 + 8)^{-\frac{3}{2}}(10x)$
$f = \dfrac{\sqrt[3]{x^2 + 7x}}{9}$	$f = \dfrac{1}{9}(x^2 + 7x)^{\frac{1}{3}}$ $f' = \dfrac{1}{27}(x^2 + 7x)^{\frac{-2}{3}}(2x + 7)$

$f = \sqrt[5]{(4x + 7)^3 + 5x}$

$f = ((4x + 7)^3 + 5x)^{\frac{1}{5}}$

$\boxed{\text{Use Chain Rule again}}$

$f' = \dfrac{1}{5}((4x + 7)^3 + 5x)^{\frac{-4}{5}}\ [\,3(4x + 7)^2(4) + 5\,]$

$f' = \dfrac{1}{5}((4x + 7)^3 + 5x)^{\frac{-4}{5}}\ [\,12(4x + 7)^2 + 5\,]$

Product Rule

Differentiation – Product Rule

Given: $u = u(x)$ and $v = v(x)$

$$\frac{d}{dx}(u \cdot v) = u'v + uv'$$

Product Rule – Simple Example

Given: $f(x) = (2x^3)^5$ Find: $f'(x)$

$$f(x) = (2x) \cdot (5x^3)$$

$$f'(x) = [2](5x^3) + (2x)[15x^2]$$

$$f'(x) = 10x^3 + 30x^3$$

$$f'(x) = 40x^3$$

Differentiation – Product Rule -- Examples

Given: $f = u \cdot v$ Find: $f' = u'v + u v'$

$f = (5x + 2) \cdot (3x)$

$f' = (5)(3x) + (5x + 2)(3)$

$f = \dfrac{(3x - 5)}{(x + 7)} = (3x - 5)(x + 7)^{-1}$

$f' = (3)(x + 7)^{-1} + (3x - 5)(-1)(x + 7)^{-2}$

$f' = (3)(x + 7)^{-1} - (3x - 5)(x + 7)^{-2}$

$f' = \dfrac{3(x + 7) - (3x - 5)}{(x + 7)^2} = \dfrac{26}{(x + 7)^2}$

$f = (x^2 + 7) \sqrt[3]{2x + 5}$

$f = (x^2 + 7)(2x + 5)^{\frac{1}{3}}$

$f' = (2x)(2x + 5)^{\frac{1}{3}} + (x^2 + 7)\left[\frac{1}{3}(2x + 5)^{\frac{-2}{3}}(2)\right]$

$f' = (2x)(2x + 5)^{\frac{1}{3}} + (x^2 + 7)\left[\frac{2}{3}(2x + 5)^{\frac{-2}{3}}\right]$

Differentiation – Chain and Product Rule -- Example

Given: $f(x) = e^{\frac{2}{x}}$

Find: 1st and 2nd derivatives

f'	$f(x) = e^{2x^{-1}}$	$u = 2x^{-1}$
	$f'(x) = e^{2x^{-1}}(-2x^{-2})$	$\frac{du}{dx} = -2x^{-2}$
	$f'(x) = -2\,e^{2x^{-1}}\,x^{-2}$	**Chain Rule**
f''	$f' = -2\left(e^{2x^{-1}}\right)(x^{-2})$	$u = e^{2x^{-1}}$
	$f' = -2(u)(v)$	$u' = e^{2x^{-1}}(-2x^{-2})$
		$v = x^{-2}$
		$v' = -2x^{-3}$
	$f''(x) = -2\,[\,uv' + u'v\,]$ **Product Rule**	

$$f''(x) = -2\left[\left(e^{2x^{-1}}\right)(-2x^{-3}) + \left(-2\,e^{2x^{-1}}x^{-2}\right)(x^{-2})\right]$$

$$f''(x) = 4\left[\left(e^{2x^{-1}}\right)(x^{-3}) + \left(e^{2x^{-1}}x^{-2}\right)(x^{-2})\right]$$

$$f''(x) = 4\left(e^{2x^{-1}}\right)\left[x^{-3} + x^{-4}\right]$$

Quotient Rule

Differentiation – Quotient Rule

Given: $u = u(x)$ and $v = v(x)$

$$\frac{d}{dx}\left(\frac{u}{v}\right) = \frac{v\,u' - v'u}{v^2}$$

Quotient Rule – Simple Example

Given: $f(x) = \dfrac{2x}{x^3 + 5}$ Find: $f'(x)$

$$f'(x) = \frac{(x^3+5)[\,2\,] - [3x^2](2x)}{(x^3+5)^2}$$

$$f'(x) = \frac{2x^3 + 10 - 6x^3}{(x^3+5)^2}$$

$$f'(x) = \frac{10 - 4x^3}{(x^3+5)^2}$$

$$f'(x) = -\frac{2(2x^3 - 5)}{(x^3+5)^2}$$

Differentiation – Quotient Rule -- Examples

Given: $f = \dfrac{u}{v}$ Find: $f' = \dfrac{v\,u' - v'u}{v^2}$

$f = \dfrac{3x}{(x^2 + 1)}$

$f' = \dfrac{(x^2 + 1)(3) - (3x)(2x)}{(x^2 + 1)^2} = \dfrac{3 - 3x^2}{(x^2 + 1)^2}$

$f = \dfrac{(3x - 5)}{(x + 7)}$ $x \neq -7$

$f' = \dfrac{(x + 7)(3) - (3x - 5)(1)}{(x + 7)^2}$

$f' = \dfrac{3x + 21 - (3x - 5)}{(x + 7)^2} = \dfrac{26}{(x + 7)^2}$

$f = \dfrac{(x - 5)}{(x^2 + 7)}$

$f' = \dfrac{(x^2 + 7)(1) - (x - 5)(2x)}{(x^2 + 7)^2}$

$f' = \dfrac{x^2 + 7 - (2x^2 - 10x)}{(x^2 + 7)^2} = \dfrac{-x^2 + 10x + 7}{(x^2 + 7)^2}$

Logs and Exponents

Differentiation – Logs & Exponent Rules

$\dfrac{d}{dx}[\ln x]$	$=$	$\dfrac{1}{x}$
$\dfrac{d}{dx}[\ln u]$	$=$	$\dfrac{1}{u}u' = \dfrac{u'}{u}$
$\dfrac{d}{dx}[\log_a u]$	$=$	$\dfrac{u'}{(\ln a)\,u}$
$\dfrac{d}{dx}[e^x]$	$=$	e^x
$\dfrac{d}{dx}[e^u]$	$=$	$e^u u'$

Logs & Exponent Rules – Simple Example

$$f = 2 \cdot \ln x$$
$$f' = 2\left(\frac{1}{x}\right) = \left(\frac{2}{x}\right)$$

$$f = e^x$$
$$f' = e^x$$

Differentiation – Logs & Exponent -- Examples
Given: $f(x)$ Find: $f'(x)$

$f = 5 \cdot \ln x$

$f' = 5\left(\frac{1}{x}\right) = \left(\frac{5}{x}\right)$

$f = \ln(x^3 + 5x)$

$f' = \frac{1}{(x^3 + 5x)} \cdot (3x^2 + 5) = \frac{3x^2 + 5}{x^3 + 5x}$

$f = e^x$

$f' = e^x$

$f = e^{(2x^7 + 3x + 2)}$

$f' = e^{(2x^7 + 3x + 2)}(14x^6 + 3)$

$f = \log x^2$

$f' = \frac{2x}{(\ln 10)\, x^2} = \frac{2}{(\ln 10)\, x}$

<u>Trig Functions</u>

Differentiation – Trig Functions		
$\frac{d}{dx}[\sin x]$	$=$	$\cos x$
$\frac{d}{dx}[\cos x]$	$=$	$-\sin x$
$\frac{d}{dx}[\tan x]$	$=$	$\sec^2 x$
$\frac{d}{dx}[\csc x]$	$=$	$-\csc x \cdot \cot x$
$\frac{d}{dx}[\sec x]$	$=$	$\sec x \cdot \tan x$
$\frac{d}{dx}[\cot x]$	$=$	$\csc^2 x$

Trig Functions – Simple Example	
Given: $f = 5 \cdot \cos x$	Find: f'
$f' = 5(-\sin x) = -5\sin x$	

Differentiation – Trig Examples
Given: $f(x)$ Find: $f'(x)$

$f = 5 \cdot \cos(2x)$

$f' = 5(-\sin(2x)(2)) \ = \ -10\sin 2x$

$f = 5 \cdot \cos(x^3)$

$f' = 5(-\sin(x^3)(3x^2)) \ = \ -10x^2 \sin(x^3)$

$f = 5 \cdot \sin^3 x$

$f' = 15(\sin x)^2(\cos x) \ = \ 15(\sin^2 x)(\cos x)$

$f = \tan(5x^4 + x^3 + 1)$

$f' = \sec^2(5x^4 + x^3 + 1) \cdot (20x^3 + 3x^2)$

$f = \sec(2x^5)$

$f' = \sec(2x^5) \cdot \tan(2x^5) \cdot (10x^4)$

$f = 5 \cdot \cot(2x)$

$f' = 5(\csc^2(2x) \cdot (2)) \ = \ 10\csc^2(2x)$

Note: $\sin^n x \ = \ (\sin x)^n \ \neq \ \sin(x^n)$

<u>Inverse Trig Functions</u>

Differentiation – Inverse Trig Functions

$$\frac{d}{dx}[\sin^{-1} x] \quad = \quad \frac{1}{\sqrt{1-x^2}}$$

$$\frac{d}{dx}[\cos^{-1} x] \quad = \quad -\frac{1}{\sqrt{1-x^2}}$$

$$\frac{d}{dx}[\tan^{-1} x] \quad = \quad \frac{1}{1+x^2}$$

$$\frac{d}{dx}[\csc^{-1} x] \quad = \quad -\frac{1}{x\sqrt{x^2-1}}$$

$$\frac{d}{dx}[\sec^{-1} x] \quad = \quad \frac{1}{x\sqrt{x^2-1}}$$

$$\frac{d}{dx}[\cot^{-1} x] \quad = \quad -\frac{1}{1+x^2}$$

Inverse Trig Functions – Simple Example

Given: $f = 5\arcsin x$ Find: f'

$$f' \quad = \quad \frac{5}{\sqrt{1-x^2}}$$

Differentiation – Inverse Trig Function Examples

Given: $f(x)$	Find: $f'(x)$

$f = \sin^{-1}(5x)$

$$f' = \frac{1}{\sqrt{1-(5x)^2}} \cdot (5) = \frac{5}{\sqrt{1-25x^2}}$$

$f = \cos^{-1}(x^2 - 5)$

$$f' = -\frac{1}{\sqrt{1-(x^2-5)^2}} \cdot (2x)$$

$$f' = \frac{-2x}{\sqrt{1-[x^4-10x^2+25]}} = \frac{-2x}{\sqrt{-x^4+10x^2-24}}$$

$f = 5\tan^{-1}(9x)$

$$f' = \frac{5}{1+(9x)^2} \cdot (9) = \frac{45}{1+81x^2}$$

$f = \tan^{-1}(\sqrt{x})$

$$f' = \frac{1}{1+\sqrt{x}^2} \left(\frac{1}{2} x^{-\frac{1}{2}}\right) = \frac{1}{2(1+x)\sqrt{x}}$$

Differentiation – Relative Extrema Example

Given: $f(x) = \arctan x - \arctan(x - 4)$

Find: Relative extrema.

$f'(x) =$ Slope of tangent line.	$f'(x) = \dfrac{1}{1 + x^2} - \dfrac{1}{1 + (x-4)^2}$
Extrema when <u>slope</u> of tangent line $= 0$	$0 = \dfrac{1}{1 + x^2} - \dfrac{1}{1 + (x-4)^2}$ $\dfrac{1}{1 + x^2} = \dfrac{1}{1 + (x-4)^2}$ $1 + x^2 = 1 + (x - 4)^2$ $x^2 = (x - 4)^2$ $x^2 = x^2 - 8x + 16$ $0 = -8x + 16$
Extrema when $x = 2$	$0 = x - 2$ $x = 2$

Continued …

Differentiation – Relative Extrema Example (Cont.)

Given: $f(x) = \arctan x - \arctan(x - 4)$

Find: Relative extrema.

Previously Found	Extrema when $x = 2$

$f(2) = \arctan(2) - \arctan(2 - 4)$

$f(2) = 1.107 - (-1.107) \approx 2.21$

Relative Extrema: $(x, y) = (2,\ 2.21)$

This point is either a relative maximum
or a relative minimum.

The relative extrema is: 2.21
which occurs at: $x = 2$

Implicit Differentiation

Implicit Differentiation

If: $y = f(x, y)$

Then: $\frac{d}{dx}[y] = \frac{d}{dx}[f(x,y)]$

In other words ...

If it is difficult to separate the x's and y's, differentiate both sides of the equation with respect to x, using the chain rule. Remember, y is a function of x.

Implicit Differentiation – Simple Example

Given: $x + y^3 = 5$ Find: y'

$$\frac{d}{dx}[x + y^3] = \frac{d}{dx}[5]$$

$$\frac{d}{dx}[x] + \frac{d}{dx}[y^3] = 0$$

$$1 + 3y^2(y') = 0$$

$$y' = -\frac{1}{3y^2} = \frac{-1}{3\left(\sqrt[3]{5-x}\right)^2} = \frac{-1}{3(5-x)^{\frac{2}{3}}}$$

Implicit Differentiation -- Ex. 1

Given: $x^2 + y^2 = 9$ Find: y'

$x^2 + y^2$	$=$	9
$\dfrac{d}{dx}[x^2 + y^2]$	$=$	$\dfrac{d}{dx}[9]$
$\dfrac{d}{dx}x^2 + \dfrac{d}{dx}y^2$	$=$	0
$2x + 2y(y')$	$=$	0
$2y(y')$	$=$	$-2x$
y'	$=$	$-\dfrac{2x}{2y}$
y'	$=$	$-\dfrac{x}{y}$

y' represents the slope of a tangent line,
at any point (x, y) on the given circle.

Implicit Differentiation -- Ex. 2

Given: $x^3 + y^3 = 5xy$ Find: y'

$$x^3 + y^3 \quad = \quad 5xy$$

$$\frac{d}{dx}[\,x^3 + y^3\,] \quad = \quad \frac{d}{dx}[\,5xy\,]$$

$$\frac{d}{dx}x^3 + \frac{d}{dx}y^3 \quad = \quad 5 \cdot \frac{d}{dx}[\,x \cdot y\,]$$

$$3x^2 + 3y^2(y') \quad = \quad 5 \cdot \{\,(1)y + xy'\,\}$$

$$3x^2 + 3y^2\,y' \quad = \quad 5y + 5xy'$$

$$3y^2\,y' - 5xy' \quad = \quad 5y - 3x^2$$

$$y' \cdot (3y^2 - 5x) \quad = \quad 5y - 3x^2$$

$$y' \quad = \quad \frac{5y - 3x^2}{3y^2 - 5x}$$

y' represents the slope of a tangent line
at any point (x, y) on the given curve.

Implicit Differentiation -- Ex. 3
Given: $\sin(x + y) = y^3 \cos x$ Find: y'

$\sin(x + y)$	$=$	$y^3 \cos x$
$\dfrac{d}{dx}[\sin(x + y)]$	$=$	$\dfrac{d}{dx}[y^3 \cos x]$
$\cos(x + y)[1 + y']$	$=$	$[3y^2 y']\cos x$ $+ y^3[-\sin x]$
$\cos(x + y)$ $+ \cos(x + y)y'$	$=$	$3y^2 y' \cos x$ $- y^3 \sin x$
$y' \cdot$ $\{\cos(x + y) - 3y^2 \cos x\}$	$=$	$-\cos(x + y)$ $- y^3 \sin x$
$y' \quad =$		$\dfrac{-\cos(x + y) - y^3 \sin x}{\cos(x + y) - 3y^2 \cos x}$
$y' \quad =$		$-\dfrac{\cos(x + y) + y^3 \sin x}{\cos(x + y) - 3y^2 \cos x}$

Implicit Differentiation -- Ex. 4a

Given: $x^2 + x \arctan y = y - 1$

Find: Eqn. of tangent line at point $(x, y) = \left(-\frac{\pi}{4}, 1\right)$

Use implicit differentiation.

$x^2 + x \arctan y = y - 1$

$\frac{d}{dx}\left[x^2 + x \arctan y\right] = \frac{d}{dx}\left[y - 1\right]$ Chain Rule

$2x + (1)(\arctan y) + (x)\left(\frac{1}{1+y^2}(y')\right) = y' - 0$

$2x + \arctan y + \frac{x\,y'}{1+y^2} = y'$

$2x + \arctan y = y' - \frac{x\,y'}{1+y^2}$

$2x + \arctan y = y'\left(1 - \frac{x}{1+y^2}\right)$

$2x + \arctan y = y'\left(\frac{1+y^2 - x}{1+y^2}\right)$

$y' = \frac{(2x + \arctan y)(1+y^2)}{1+y^2 - x}$

Implicit Differentiation -- Ex. 4b

Given: $x^2 + x \arctan y = y - 1$

Find: Eqn. of tangent line at point $(x, y) = \left(-\frac{\pi}{4}, 1\right)$

Use implicit differentiation.

Previously Found	$y' = \dfrac{(2x + \arctan y)(1 + y^2)}{1 + y^2 - x}$
Slope of tangent line at point $(x, y) = \left(-\frac{\pi}{4}, 1\right)$	$y' = \text{slope of tangent line}$ $y' = \dfrac{(2x + \arctan y)(1 + y^2)}{1 + y^2 - x}$ $y' = \dfrac{\left(-\frac{\pi}{2} + \frac{\pi}{4}\right)(1 + 1)}{1 + 1 + \frac{\pi}{4}} = \dfrac{\left(-\pi + \frac{\pi}{2}\right)}{2 + \frac{\pi}{4}}$ $y' = \dfrac{\left(-\frac{\pi}{2}\right)}{\left(\frac{8 + \pi}{4}\right)} = \dfrac{-2\pi}{8 + \pi}$ $\boxed{m = \text{Slope}}$
Equation of tangent line at $\left(-\frac{\pi}{4}, 1\right)$	$\Delta y = m\,\Delta x$ $\boxed{\begin{array}{l}\text{Recall:}\\ m = \frac{\Delta y}{\Delta x}\end{array}}$ $(y - 1) = m\left(x + \frac{\pi}{4}\right)$ $(y - 1) = \left(\dfrac{-2\pi}{8 + \pi}\right) \cdot \left(x + \frac{\pi}{4}\right)$

Equation of Tangent Line

Equation of a Tangent Line – General Process

Given the equation of a curve $y = f(x)$

Find the equation of a tangent line at $x = a$.

The process is outlined below.

Find $f'(x)$	$f'(x) =$ Derivative of $f(x)$ $f'(x) =$ Slope of tangent line.
Find $m = f'(a)$	$f'(a) =$ Slope of the tangent line at $x = a$.
Find the point. $(x, y) = (a, f(a))$	This is point where the tangent line touches the curve.
Find the y-intercept.	Use the slope (m) and the point $(x, y) = (a, f(a))$ to find the y-intercept $= b$.
Equation of tangent line.	$y = mx + b$

Equation of a Tangent Line – Ex. 1
Given: $f(x) = x^2$ Find: Equation of tangent line at $x = 1$

Find $f'(x)$	$f(x) = x^2$ $f'(x) = 2x$
Find $m = f'(a)$	$f'(1) = 2(1) = 2$ Slope of tangent line $= 2$
Find the point. $(x, y) = (a, f(a))$	$y = f(1) = 1$ The point is $(x, y) = (1, 1)$
Find the y-intercept.	$y = mx + b$ $y = 2x + b$ $(1, 1) \rightarrow 1 = 2(1) + b$ $-1 = b$
Equation of tangent line.	$y = mx + b$ $y = 2x - 1$

Equation of a Tangent Line – Ex. 2

Given: $f(x) = \sqrt{x}$

Find: Equation of tangent line at $x = 4$

Find $f'(x)$	$f(x) = \sqrt{x} = x^{1/2}$ $f'(x) = \frac{1}{2}x^{-\frac{1}{2}} = \frac{1}{2\sqrt{x}}$
Find $m = f'(a)$	$f'(4) = \frac{1}{2\sqrt{4}} = \frac{1}{4}$ Slope of tangent line $= \frac{1}{4}$
Find the point. $(x, y) = (a, f(a))$	$y = f(4) = 2$ The point is $(x, y) = (4, 2)$
Find the y-intercept.	$y = \left(\frac{1}{4}\right)x + b$ $(4, 2) \rightarrow 2 = \left(\frac{1}{4}\right)4 + b$ $1 = b$
Equation of tangent line.	$y = mx + b$ $y = \frac{1}{4}x + 1$

Equation of a Tangent Line – Ex. 3

Graph the tangent lines for previous examples

$f(x) = x^2$
At $x = 1$

Equation of
Tangent Line:
$y = 2x - 1$

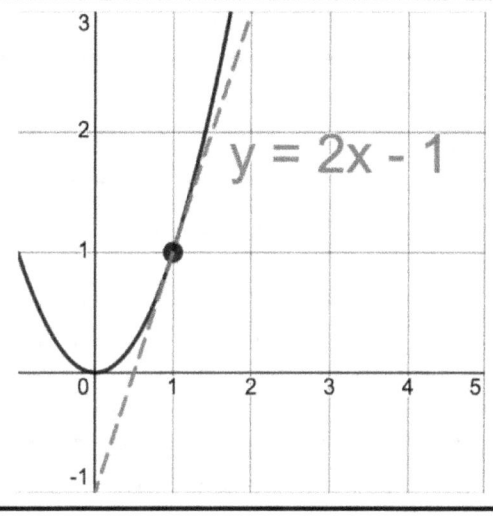

$f(x) = \sqrt{x}$
At $x = 4$

Equation of
Tangent Line:
$y = \left(\frac{1}{4}\right)x + 1$

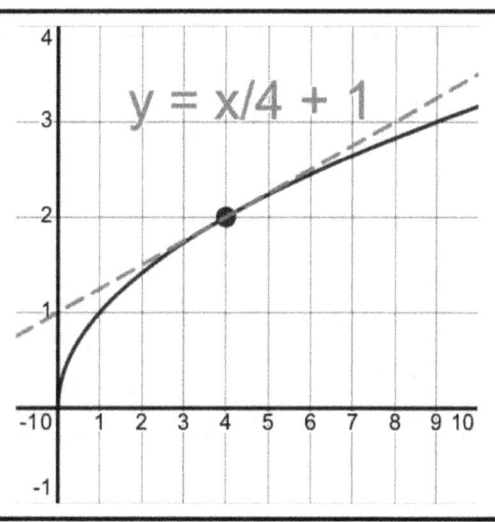

1 Point and Tangent – Quadratic Eqn. – Ex. 4

Find the quadratic equation that
- Goes through the point $(0,6)$
- Is tangent to $y = 3x - 3$ at $(1,0)$

Quad. Eqn.	$y = ax^2 + bx + c$
$(x, y) = (0, 6)$	$6 = a(0)^2 + b(0) + c$ $6 = c$
$(x, y) = (1, 0)$	$0 = a(1)^2 + b(1) + c$ $0 = a + b + 6 \qquad \rightarrow \quad b = -6 - a$
Find y'	$y = ax^2 + bx + c$ $y' = 2ax + b$
$y' = 3$ At point $(1, 0)$	$y' = 2ax + b$ $3 = 2a(1) + b$ $3 = 2a(1) + [-6 - a]$ $9 = a \qquad \rightarrow \quad b = -6 - 9 = -15$
Equation	$y = ax^2 + bx + c$ $y = 9x^2 - 15x + 6$

Linear Approximation

Linear Approximation

Suppose it is difficult to evaluate a function at a value of x. To approximate the value of the function, you could evaluate the function at point, near to x, that is easier to evaluate. For example, finding the square root of 4.05 is difficult. But, finding the square root of 4 is easy!

Using "Linear Approximation" is a more accurate way to approximate a function near a particular value of x.

Linear Approximation of $f(x)$
For points near $f(a)$

$$L(x) = mx + b$$

$$L(x) = f'(a)(\Delta x) + f(a)$$

$$L(x) = f'(a)(x - a) + f(a)$$

Note: $L(x) \approx f(a)$

Adding $f'(a)(\Delta x)$ makes it more accurate.

Linear Approximation. – Ex. 1

Find $\sqrt{4.05}$

Use: $L(x) = f'(a)(x - a) + f(a)$

Identify The parts	$f(x) = \sqrt{x} = x^{1/2}$ $f'(x) = \frac{1}{2} x^{-1/2} = \frac{1}{2\sqrt{x}}$ $x = 4.05$, $\quad a = 4$
Evaluate f & f' at a	$f(4) = \sqrt{4} = 2$ $f'(4) = \frac{1}{2\sqrt{4}} = \frac{1}{4}$
Linear Approx.	$L(x) = f'(a)(x - a) + f(a)$ $L(4.05) = f'(4)(0.05) + f(4)$ $L(4.05) = \left(\frac{1}{4}\right)(0.05) + 2$ $L(4.05) = 0.0125 + 2 = 2.0125$

Linear Approximation. – Ex. 2	
Find: $\sqrt{3.98}$	
Use: $L(x) = f'(a)(x-a) + f(a)$	

Identify The parts	$f(x) = \sqrt{x} = x^{1/2}$ $f'(x) = \frac{1}{2}x^{-1/2} = \frac{1}{2\sqrt{x}}$ $x = 3.98$, $a = 4$
Evaluate f & f' at a	$f(4) = \sqrt{4} = 2$ $f'(4) = \frac{1}{2\sqrt{4}} = \frac{1}{4}$
Linear Approx.	$L(x) = f'(a)(x-a) + f(a)$ $L(3.98) = f'(4)(-0.02) + f(4)$ $L(3.98) = \left(\frac{1}{4}\right)(-0.02) + 2$ $L(3.98) = -0.005 + 2 = 1.995$

Differentials and Error

Differentials $\dfrac{\Delta y}{\Delta x} \approx \dfrac{dy}{dx}$

Point	Description
P	On curve and on tangent line At some point (x, y)
Q	On curve at point $(x + \Delta x,\ y + \Delta y)$
R	On tangent at point $(x + dx,\ y + dy)$

Differential $= dy = f'(x)\,dx$
Because: $\quad \dfrac{dy}{dx} \ = \ f'(x)$ $\qquad dy \ = \ f'(x)\,dx$

Slope of Tangent Line	$\dfrac{dy}{dx} \ = \ f'(x)$
Slope of $\overline{PQ}$	$\dfrac{\Delta y}{\Delta x} \ = \ \dfrac{Change\ in\ y}{Change\ in\ x}$
$\overline{PQ}$	Line segment on the curve from P to Q.

Differentials $\quad \dfrac{\Delta y}{\Delta x} \approx \dfrac{dy}{dx} \quad$ **(on one graph)**

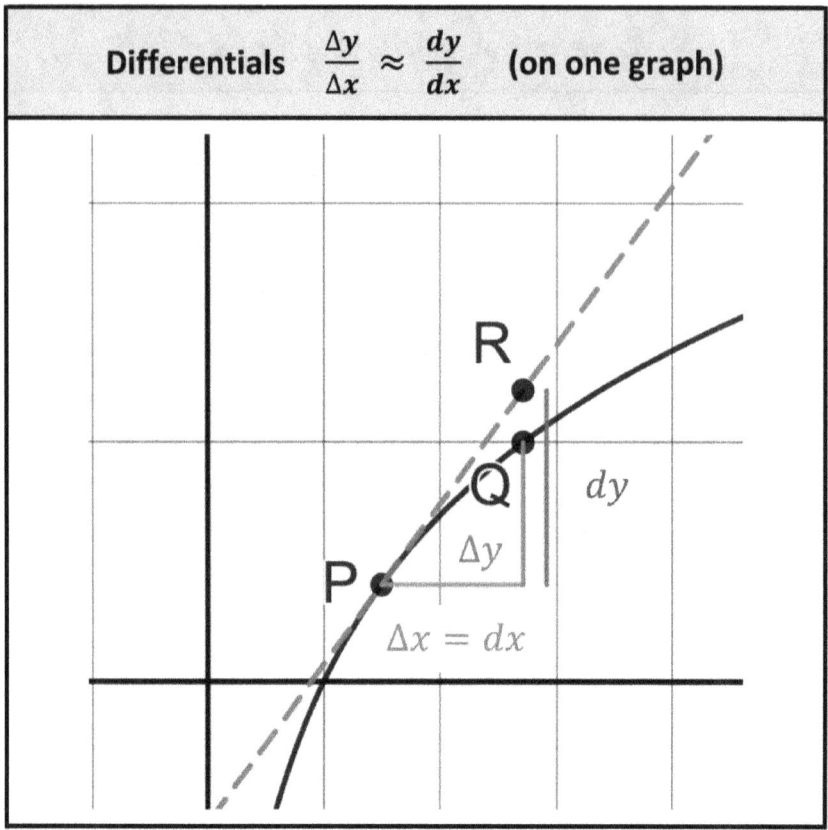

Point	Description
P	On curve and on tangent line At some point (x, y)
Q	On curve at point $(x + \Delta x,\ y + \Delta y)$
R	On tangent at point $(x + dx,\ y + dy)$

Differential Error Propagation -- Ex. 1

Given: Sphere radius $= r = 23 \pm 0.05$ cm

Find: Volume of sphere and max. error.

Volume Eqn.	$V = \frac{4}{3}\pi r^3$
Take derivative of both sides	$\frac{d}{dr}[V] = \frac{d}{dr}\left[\frac{4}{3}\pi r^3\right]$ $\frac{dV}{dr} = \frac{4\pi}{3}[3r^2] = 4\pi r^2$ $dV = 4\pi r^2\, dr$
Max. Error $\approx 332\ cm^3$	$dV = 4\pi(23)^2\,(.05) \approx 332$
Relative Error $\approx 0.65\%$	$\frac{dV}{V} = \frac{4\pi r^2\, dr}{\frac{4}{3}\pi r^3} = 3\frac{dr}{r}$ $\frac{dV}{V} = 3\frac{0.05}{23} \approx 0.0065$ $\approx 0.65\%$

Applications of Derivatives

Maximum and Minimum Values

Local Maximum & Minimum

The local maximum is a value that is larger than near-by values. A local minimum is a value that is smaller than near-by values. There may be many local max and min within an interval.

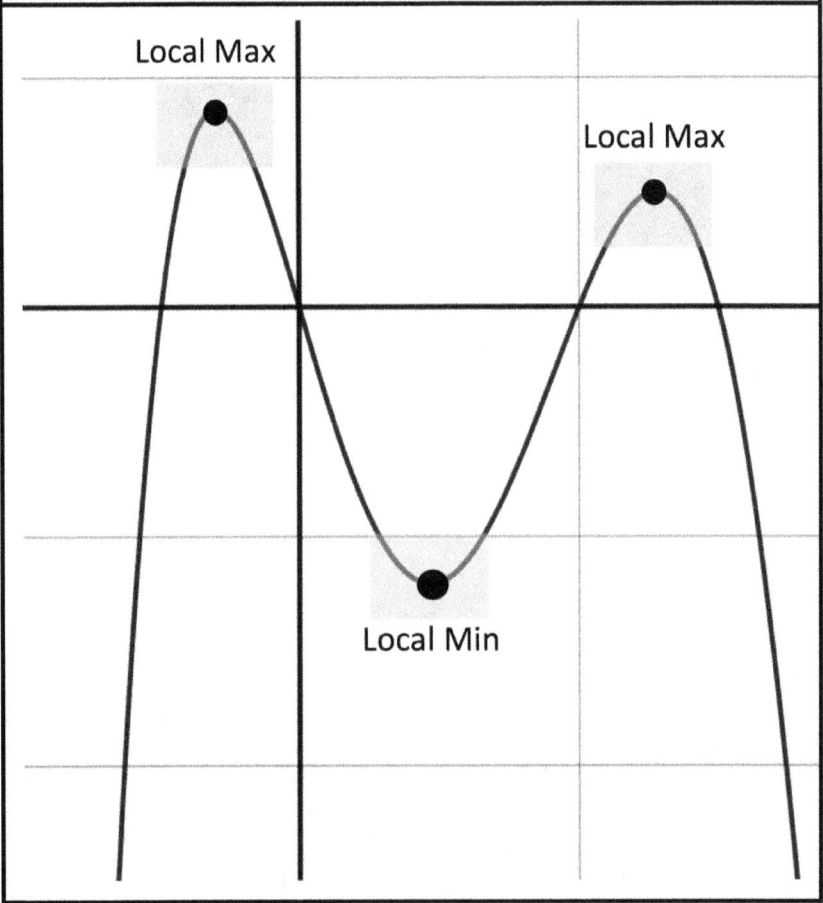

Absolute Maximum & Minimum

The absolute maximum is the largest value of the function within the domain. The absolute minimum is the smallest value of the function within the domain. For closed intervals, there can be only <u>one</u> absolute max and <u>one</u> absolute min.

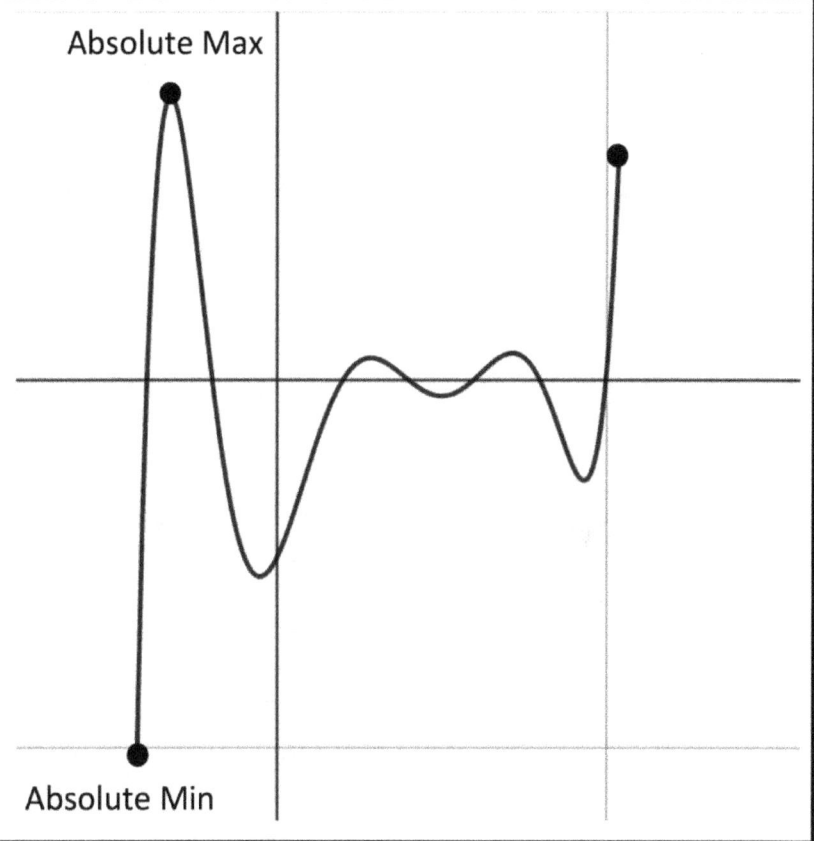

Absolute Maximum & Minimum (another graph)

For closed intervals, there can be only <u>one</u> absolute max and <u>one</u> absolute min.

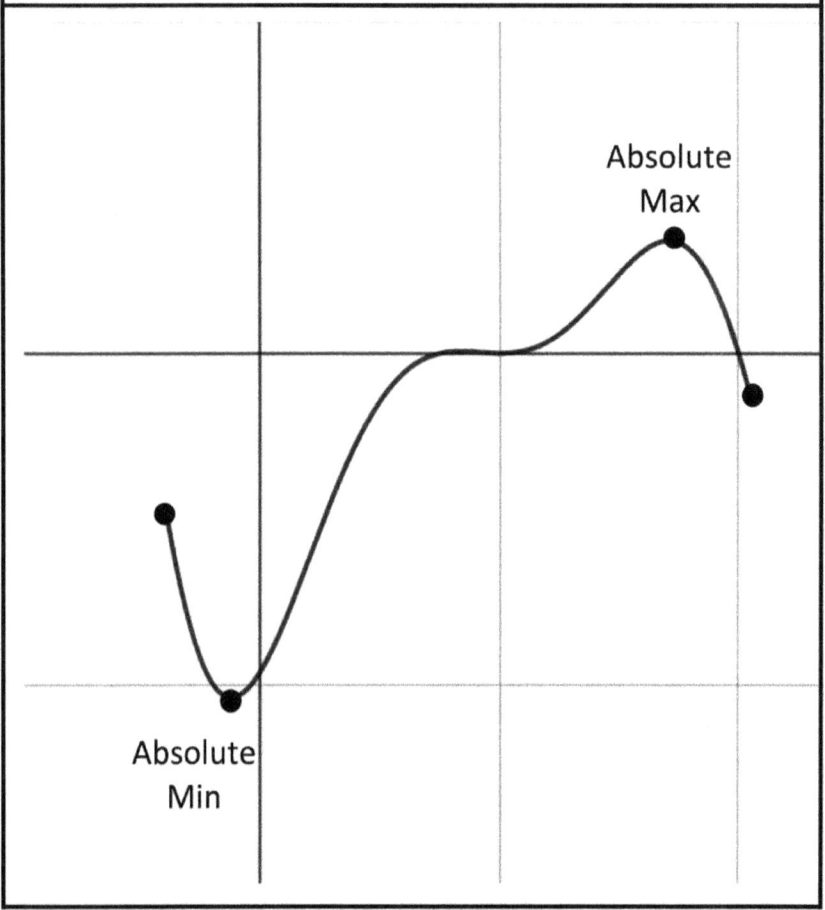

Extreme Value Theorem

If

f is continuous

on a closed interval $[a, b]$

Then

f has an absolute max. value $f(c)$

and an absolute min. value $f(d)$

at some numbers $c \; and \; d$

in the interval $[a, b]$.

Note: An extreme value may or may not

occur at the boundaries.

At Max & Min Points
Tangents have a Slope $= 0$

Extreme values (max and min)
on a smooth, continuous curve,
are where the slope of the tangent line is zero.

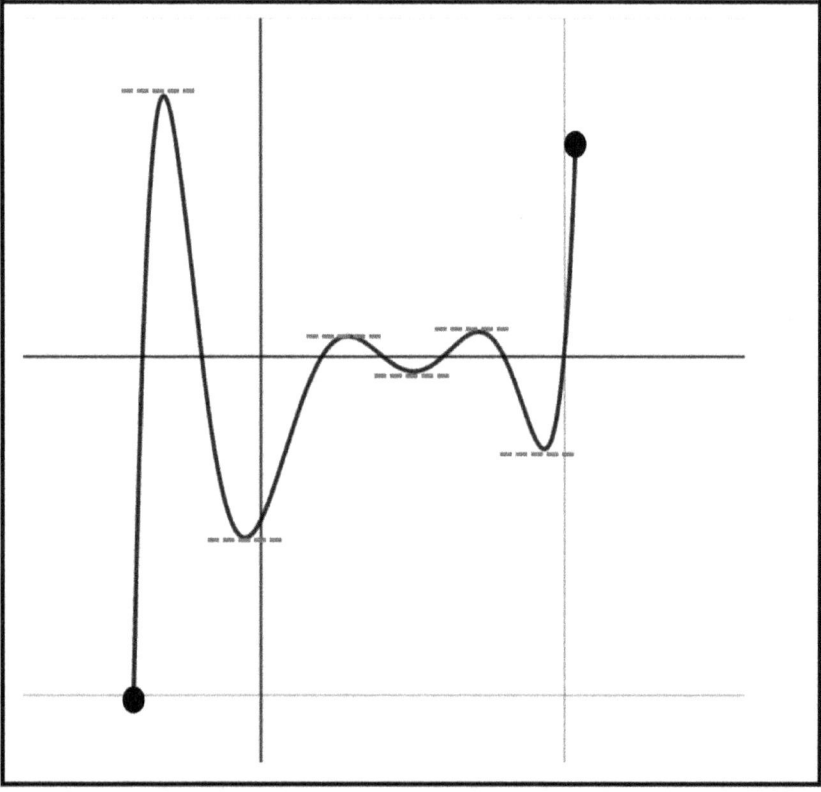

Critical Values

To find the maximum and minimum values
for a function, find the critical values.
Then, evaluate the function at those critical values.

A critical number of function f

is a number c in the domain of f

such that:

$f'(c) = 0$ or $f'(c)$ does not exist.

Recall: $f'(c)$ is the slope of the tangent line at $x = c$

Absolute Max & Min Values – Ex. 1

Find max. & min. for

$$f(x) = x^3 + x^2 - 4x$$

on $[-3, 2]$

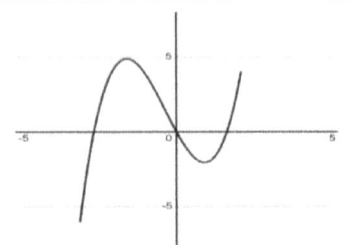

Find derivative of f	$f = x^3 + x^2 - 4x$ $f' = 3x^2 + 2x - 4$
Find critical values. Set derivative = 0 Quadratic formula: $x = \dfrac{-b \pm \sqrt{b^2 - 4ac}}{2a}$	$3x^2 + 2x - 4 = 0$ $x = \dfrac{-2 \pm \sqrt{4 - 4(3)(-4)}}{2(3)}$ $x = \dfrac{-2 \pm \sqrt{52}}{6}$ $x \approx 0.87, \; -1.5$
Evaluate <u>function</u> at boundaries and at critical values, within the boundaries. Then, select MAX and MIN.	$f(-3) = -6 \qquad$ MIN $f(2) = 4$ $f(0.87) = -2.06$ $f(-1.5) = 4.88 \qquad$ MAX

Absolute Max & Min Values − Ex. 2

Find max. & min. for

$$f = \frac{1}{4}x^4 - \frac{5}{3}x^3 + 3x^2 - 4$$

On $[-1, 4]$

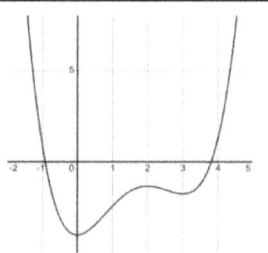

Find derivative of f	$f = \frac{1}{4}x^4 - \frac{5}{3}x^3 + 3x^2 - 4$ $f' = x^3 - 5x^2 + 6x$
Find critical values. Set $f' = 0$	$x^3 - 5x^2 + 6x = 0$ $x(x-2)(x-3) = 0$ $x = 0, 2, 3$
Evaluate <u>function</u> at boundaries and at critical values. Then select MAX & MIN.	$f(-1) \approx .92$ $f(4) \approx 1.3 \qquad$ MAX $f(0) = -4 \qquad$ MIN $f(2) \approx -1.3$ $f(3) \approx -1.8$

Absolute Max & Min Values — Ex. 3

Find max. & min. for

$$f = \frac{1}{4}x^4 - \frac{5}{3}x^3 + 3x^2 - 4$$

on $[\,1, 4\,]$

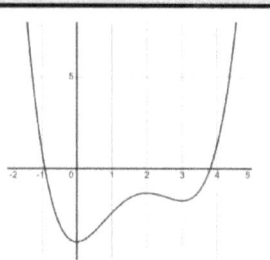

Find f'	$f = \frac{1}{4}x^4 - \frac{5}{3}x^3 + 3x^2 - 4$
	$f' = x^3 - 5x^2 + 6x$
Find critical values. Set $f' = 0$	$x^3 - 5x^2 + 6x = 0$ $x(x - 2)(x - 3) = 0$ $x = 0, 2, 3$
Evaluate <u>function</u> at boundaries and at critical values. Select MAX & MIN. Note: 0 not in $[\,1, 4\,]$	$f(1) \approx -2.4$ MIN $f(4) \approx 1.3$ MAX $f(2) \approx -1.3$ $f(3) \approx -1.8$

Absolute Max & Min Values – Ex. 4

Find
max & min
for $f(x)$
on $[a, e]$

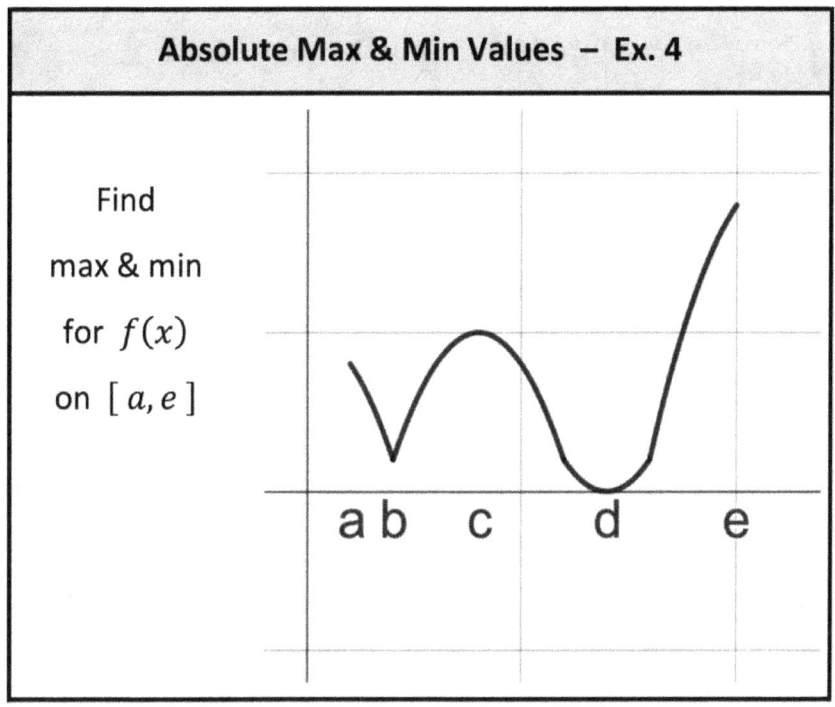

a b   c   d   e

Relative MIN	$f(b), f(d)$
Relative MAX	$f(c)$
Absolute MIN	$f(d)$
Absolute MAX	$f(e)$

Absolute Max & Min Values — Ex. 5

Find
max & min
for $f(x)$

on $[a, g)$

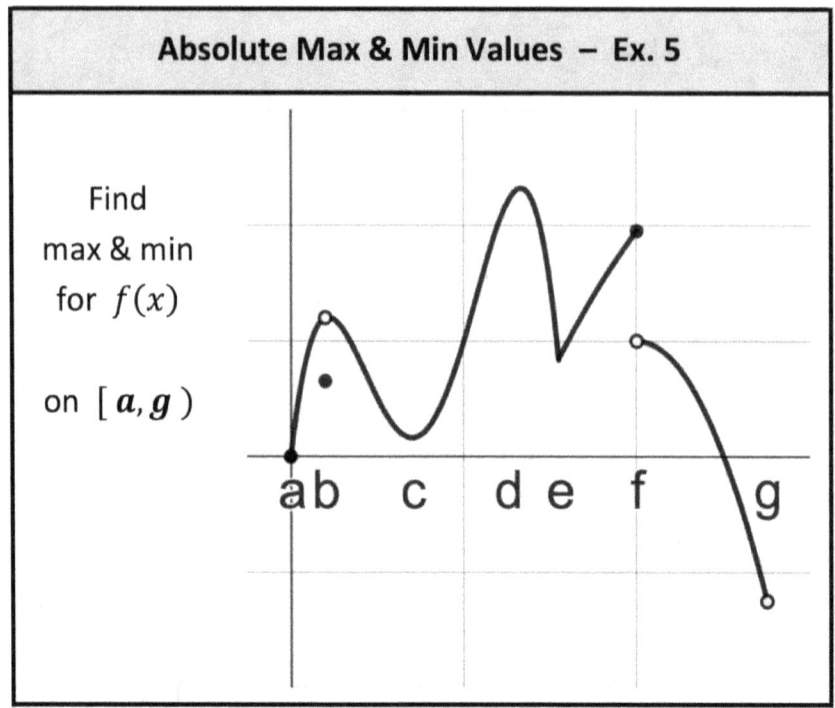

a b   c   d e   f   g

Relative MIN	$f(b), f(c), f(e)$
Relative MAX	$f(d), f(f)$
Absolute MIN	None.
Absolute MAX	$f(d)$

Mean Value Theorem (MVT)

Rolle's Theorem

There is a number c in (a, b)

Such that: $f'(c) = 0$

If the 3 conditions are met:

1. f is continuous on $[a, b]$
2. f is differentiable on (a, b)
3. $f(a) = f(b)$

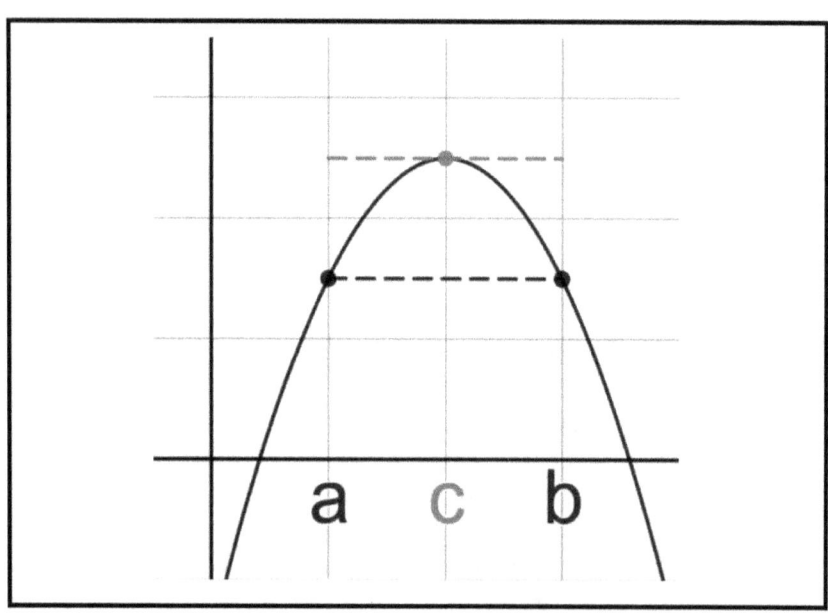

Mean Value Theorem (MVT)

There is a number c in (a, b)

such that: $f'(c) = \dfrac{f(b) - f(a)}{b - a}$

If the 2 conditions are met:

1. f is continuous on $[a, b]$
2. f is differentiable on (a, b)

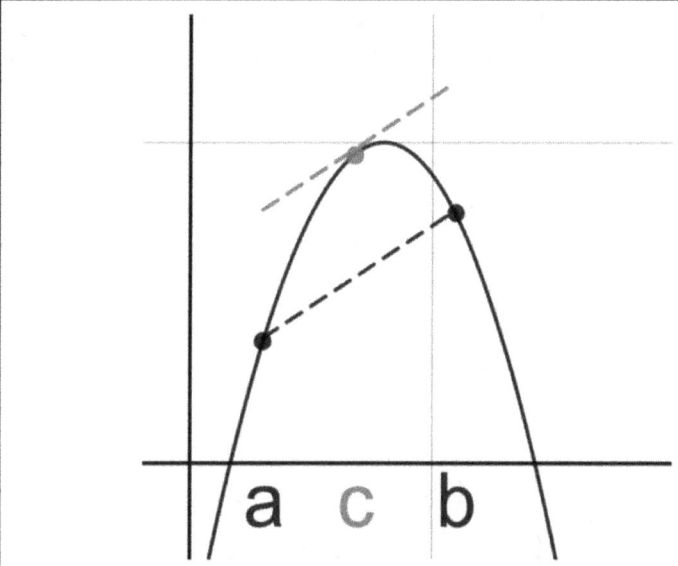

The Mean Value Theorem (MVT)
is a more general form of Role's Theorem.

Mean Value Theorem -- Ex. 1

Given: $f(x) = \frac{1}{4}x^2 + 1$

Find: Equation of tangent line

with slope: $\dfrac{f(6) - f(2)}{6 - 2} = \dfrac{10 - 2}{6 - 2} = \dfrac{8}{4} = 2$

Verify the MVT applies	$f(x)$ is a polynomial so it continuous on $[2, 6]$ and differentiable on $(2, 6)$
Find point on curve where slope of tangent $= 2$	$f'(x) = \frac{1}{2}x = 2 \quad \rightarrow \quad x = 4$ $f(4) = \frac{1}{4}(4)^2 + 1 = 5$ Point $(x, y) = (4, 5)$
Find eqn. of tangent line at point $(4, 5)$	$y = 2x + b$ $(4, 5) \rightarrow 5 = 2(4) + b$ $-3 = b$ Equation: $y = 2x - 3$

Graphing With Derivatives

Graphing With Derivatives

Derivatives provides some very useful information that makes graphing easier.

The first example is a review of graphing without using derivatives. All other examples, will demonstrate the use of derivatives to help with graphing.

$f(x)$	The curve.
$f'(x)$	1st Derivative of the function indicates the slope of the tangent line to the curve.
$f''(x)$	2nd Derivative of the function indicates the concavity of the curve. • $f'' = $ Positive $\rightarrow$ Concave Up • $f'' = $ Negative $\rightarrow$ Concave Down

Graphing Without Derivatives (Review)
Graph: $f(x) = 2x^3 - 7x^2 + 5x$

End Behavior	↙↗ Because odd degree and leading coefficient.
Zeros and Multiplicity	$f(x) = 2x^3 - 7x^2 + 5x$ $f(x) = x(2x^2 - 7x + 5)$ $f(x) = x(x - 1)(2x - 5)$ Zeros at: $x = 0, 1, 2.5$ All multiplicities $= 1$
Graph	

Graphing With f' -- Ex. 1
Graph: $f(x) = 2x^3 - 7x^2 + 5x$

Find zeroes and end behavior of f	$f = 2x^3 - 7x^2 + 5x$ $f = x(x-1)(2x-5)$ Zeros at: $x = 0, 1, 2.5$ All mult. $= 1$ End: ↙↗
Find extrema. Set $f' = 0$	$f' = 6x^2 - 14x + 5 = 0$ $x = \dfrac{14 \pm \sqrt{14^2 - 4(6)(5)}}{2(6)}$ $x \approx 0.4, \ 1.9$
Graph	

Graphing With f' and f'' -- Ex. 2

Graph: $f(x) = 2x^3 - 7x^2 + 5x$

Set $f = 0$ To find zeroes	$f = x(x - 1)(2x - 5) = 0$ $x = 0, 1, 2.5$ End: ↙↗
Set $f' = 0$ To find extrema	$f' = 6x^2 - 14x + 5 = 0$ $x = \dfrac{14 \pm \sqrt{14^2 - 4(6)(5)}}{2(6)}$ $x \approx 0.4,\ 1.9$
<u>Use</u> f'' To find concavity.	$f''(x) = 12x - 14$ $f''(0.4) = $ negative $f''(1.9) = $ positive
Graph	

Find Extrema With f' and f'' -- Ex. 3a

Find all extrema points

For: $f(x) = x^4 - 3x^2 - 2$

Find f' and f''	$f' = 4x^3 - 6x$ $f'' = 12x^2 - 6$
Set $f' = 0$ To find extrema	$f' = 4x^3 - 6x = 0$ $f' = 2x(2x^2 - 3) = 0$ $x = 0, \pm\sqrt{\frac{3}{2}} \approx 0, \pm 1.2$
Use f'' To check concavity at extrema.	$f''(0) = -6$ $f''(1.2) = 11.3$ $f''(-1.2) = 11.3$
Conclusion	Local MAX at $x = 0$ Local MIN at $x = \pm 1.2$

Find Extrema With f' and f'' -- Ex. 3b

Find all extrema points

For: $f(x) = x^4 - 3x^2 - 2$

Previous Conclusion	Local MAX at $\quad x = 0$ Local MIN at $\quad x = \pm 1.2$
Extra: Here is the graph of $f(x)$.	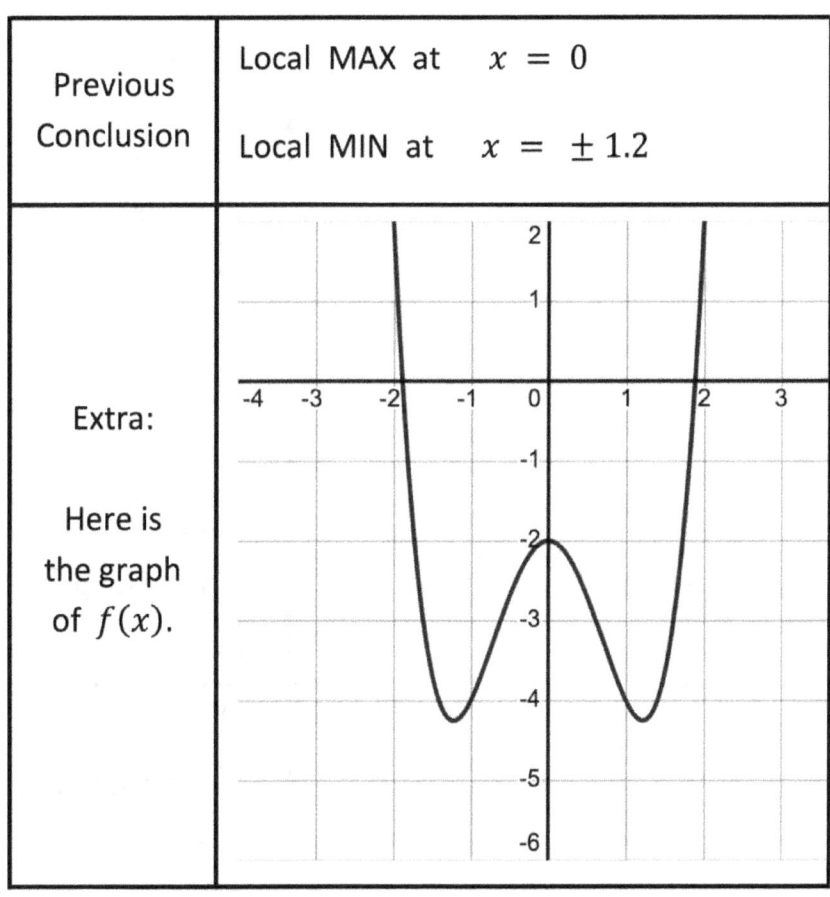

Comparison of f, f' and f''

$$f(x) = 2x^3 - 7x^2 + 5x$$
$$f'(x) = 6x^2 - 14x + 5$$
$$f''(x) = 12x - 14$$

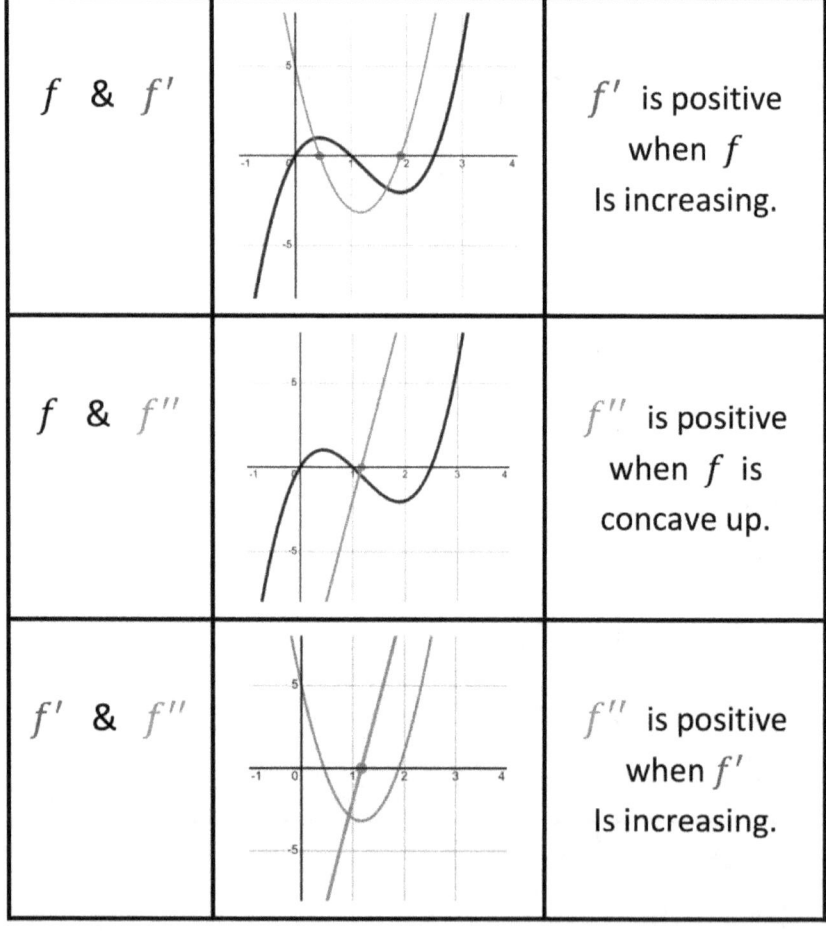

f & f'		f' is positive when f Is increasing.
f & f''		f'' is positive when f is concave up.
f' & f''		f'' is positive when f' Is increasing.

Graphing Information Summary

$f(x)$		
	$+$	Function is above x-axis
	$-$	Function is below x-axis
	0	Function crosses x-axis.

$f'(x)$		
	$+$	Function is increasing.
	$-$	Function is decreasing.
	0	Function extrema. (max or min)

$f''(x)$		
	$+$	Function is concave up.
	$-$	Function is concave down.
	0	Function has an inflection point.

Derivatives of Piece-Wise Function

Functions With Distance and Time	
Distance is a function of time. $y = f(t)$	$y = distance$ $y = f(t)$
Velocity is the change in distance over time. $v = \dfrac{\Delta y}{\Delta t}$	$v = \dfrac{dy}{dt} = f'(t)$
Acceleration is the change in velocity over time. $a = \dfrac{\Delta v}{\Delta t}$	$a = \dfrac{dv}{dt} = \dfrac{d}{dt}\left[\dfrac{dy}{dt}\right]$ $a = f''(t)$

In this section, we will look at a piece-wise function with distance and time.

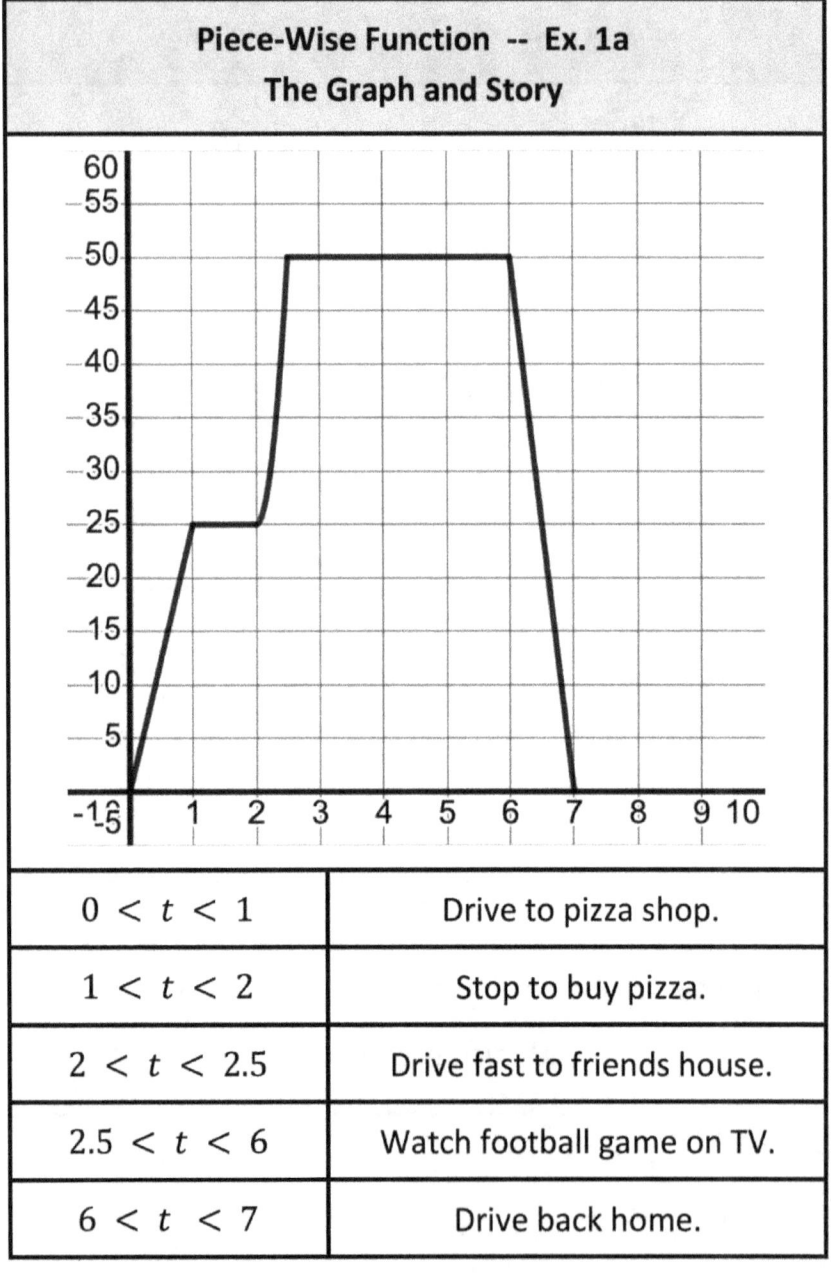

Piece-Wise Function -- Ex. 1a	
The Graph and Story	
$0 < t < 1$	Drive to pizza shop.
$1 < t < 2$	Stop to buy pizza.
$2 < t < 2.5$	Drive fast to friends house.
$2.5 < t < 6$	Watch football game on TV.
$6 < t < 7$	Drive back home.

Piece-Wise Function -- Ex. 1b

Equations for: $f,\ f',\ and\ f''$

$$f = distance$$

$$f' = velocity$$

$$f'' = acceleration$$

$$f(t) = \begin{cases} 25t & 0 < t \le 1 \\ 25 & 1 < t \le 2 \\ 100(t-2)^2 + 25 & 2 < t \le 2.5 \\ 50 & 2.5 < t \le 6 \\ -50t + 350 & 6 < t \le 7 \end{cases}$$

$$f'(t) = \begin{cases} 25 & 0 < t \le 1 \\ 0 & 1 < t \le 2 \\ 200t - 400 & 2 < t \le 2.5 \\ 0 & 2.5 < t \le 6 \\ -50 & 6 < t \le 7 \end{cases}$$

$$f''(t) = \begin{cases} 0 & 0 < t \le 1 \\ 0 & 1 < t \le 2 \\ 200 & 2 < t \le 2.5 \\ 0 & 2.5 < t \le 6 \\ 0 & 6 < t \le 7 \end{cases}$$

Piece-Wise Function -- Ex. 1c
Graphs for: $f, f', and\ f''$

$f(t)$ Distance from home.	
$f'(t)$ Velocity	
$f''(t)$ Acceleration	

L'Hospital's Rule

When to Use L'Hospital's Rule
(Pronounced "loh-pee-THAL")

When a limit results in an

indeterminate form $\left(\frac{0}{0} \ or \ \frac{\infty}{\infty} \right)$

use L'Hospital's Rule.

Example: $\displaystyle\lim_{x \to 1} \frac{\ln x}{x - 1} \ = \ \frac{\ln 1}{1 - 1} \ = \ \frac{0}{0}$

Some Indeterminate Forms

$\dfrac{0}{0}$	$\dfrac{\pm\infty}{\pm\infty}$	$\infty - \infty$
0^0	∞^0	$0 \cdot \infty$

Some Forms that are NOT Indeterminate
(They can be determined!)

$\dfrac{0}{\pm\infty} = 0$	$\dfrac{0}{\pm\infty} = 0$
$c - c = 0$	$c - c = 0$

L'Hospital's Rule

If a rational function is in an indeterminate form

$$\text{of type} \quad \frac{0}{0} \quad \text{or} \quad \frac{\infty}{\infty}$$

Then

$$\lim_{x \to a} \frac{f(x)}{g(x)} = \lim_{x \to a} \frac{f'(x)}{g'(x)}$$

L'Hospital's Rule – For One Sided Limits

If a rational function in an indeterminate form

$$\text{of type} \quad \frac{0}{0} \quad \text{or} \quad \frac{\pm\infty}{\pm\infty}$$

Then

$$\lim_{x \to a^{\pm}} \frac{f(x)}{g(x)} = \lim_{x \to a^{\pm}} \frac{f'(x)}{g'(x)}$$

L'Hospital's Rule. -- Ex. 1
Find: $\displaystyle\lim_{x \to 1} \frac{\ln x}{x-1}$

Substitution gives an indeterminant form of $\frac{0}{0}$.	$\displaystyle\lim_{x \to 1} \frac{\ln x}{x-1} = \frac{\ln(1)}{1-1} = \frac{0}{0}$
Apply LH Rule. Differentiate numerator and denominator.	$\displaystyle\lim_{x \to 1} \frac{\frac{d}{dx}(\ln x)}{\frac{d}{dx}(x-1)}$ $\displaystyle\lim_{x \to 1} \frac{\left(\frac{1}{x}\right)}{(1)}$ $\displaystyle\lim_{x \to 1} \frac{1}{x} = \frac{1}{1} = 1$

L'Hospital's Rule (Notes) -- Ex. 2a

- Sometimes the LH rule needs to be applied multiple times.

- The LH rule can only be applied to these indeterminate forms $\left(\frac{0}{0} \ or \ \frac{\infty}{\infty} \right)$

- Other indeterminate forms must be reformatted to one of the acceptable forms before the LH rule can be applied.

L'Hospital's Rule -- Ex. 2b
Find: $\lim\limits_{x \to \infty} \dfrac{e^x}{3x^2}$

Substitution gives an indeterminate form of $\dfrac{\infty}{\infty}$.	$\lim\limits_{x \to \infty} \dfrac{e^x}{3x^2} = \dfrac{\infty}{\infty}$
Apply LH Rule. But, we still have an indeterminate form.	$\lim\limits_{x \to \infty} \dfrac{\frac{d}{dx}(e^x)}{\frac{d}{dx}(3x^2)}$ $\lim\limits_{x \to \infty} \dfrac{(e^x)}{(6x)} = \dfrac{\infty}{\infty}$
Apply LH Rule again.	$\lim\limits_{x \to \infty} \dfrac{\frac{d}{dx}(e^x)}{\frac{d}{dx}(6x)}$ $\lim\limits_{x \to \infty} \dfrac{(e^x)}{(6)} = \dfrac{\infty}{6} = \infty$

LH Rule: Indeterminate Products -- Ex. 3

Find: $\displaystyle\lim_{x \to 0^+} x \cdot \ln x$

Indeterminate product.	$\displaystyle\lim_{x \to 0^+} x \cdot \ln x \;=\; 0 \cdot \infty$
Rewrite as a quotient.	$\displaystyle\lim_{x \to 0^+} \frac{\ln x}{\left(\frac{1}{x}\right)} \;=\; \frac{-\infty}{\infty}$
Quotient is indeterminate so apply the LH Rule.	$\displaystyle\lim_{x \to 0^+} \frac{\frac{d}{dx}(\ln x)}{\frac{d}{dx}(x^{-1})}$ $\displaystyle= \lim_{x \to 0^+} \frac{\left(\frac{1}{x}\right)}{(-x^{-2})}$ $\displaystyle= \lim_{x \to 0^+} \frac{-x^2}{x}$ $\displaystyle= \lim_{x \to 0^+} (-x) \;=\; 0$

LH Rule: Indeterminate Diff. -- Ex. 4a

Find: $\displaystyle \lim_{x \to 1^+} \left(\frac{1}{\ln x} - \frac{1}{x-1} \right)$

Rewrite difference as single fraction.

$$\lim_{x \to 1^+} \left(\frac{(x-1) - \ln x}{\ln x \cdot (x-1)} \right) = \frac{(0) - (0)}{(0)\cdot(0)} = \frac{0}{0}$$

Indeterminate quotient. Apply LH Rule.

$$\lim_{x \to 1^+} \frac{\frac{d}{dx}(x - 1 - \ln x)}{\frac{d}{dx}(\ln x \cdot (x-1))}$$

$$= \lim_{x \to 1^+} \frac{1 - 0 - \frac{1}{x}}{\left(\frac{1}{x}\right)(x-1) + \ln x(1)}$$

$$= \lim_{x \to 1^+} \frac{1 - \frac{1}{x}}{1 - \frac{1}{x} + \ln x}$$

$$= \frac{1 - 1}{1 - 1 + 0} = \frac{0}{0}$$

Indeterminate result. Apply LH rule again.

Continued ...

LH Rule: Indeterminate Diff. -- Ex. 4b

Find: $\displaystyle\lim_{x \to 1^+} \left(\frac{1}{\ln x} - \frac{1}{x-1} \right)$

Previously found	$\displaystyle\lim_{x \to 1^+} \frac{1 - \frac{1}{x}}{1 - \frac{1}{x} + \ln x} = \frac{0}{0}$

Apply LH Rule again ...

$$\lim_{x \to 1^+} \frac{\frac{d}{dx}\left(1 - \frac{1}{x}\right)}{\frac{d}{dx}\left(1 - \frac{1}{x} + \ln x\right)}$$

$$= \lim_{x \to 1^+} \frac{\frac{d}{dx}\left(1 - x^{-1}\right)}{\frac{d}{dx}\left(1 - x^{-1} + \ln x\right)}$$

$$= \lim_{x \to 1^+} \frac{0 + x^{-2}}{0 + x^{-2} + \frac{1}{x}}$$

$$= \frac{0 + 1}{0 + 1 + 1} = \frac{1}{2}$$

Limits of Exponential Equations (Tip) -- Ex. 5

When working with limits of exponential equations,
take the log of both sides to remove the exponent.

$$L \quad = \quad \lim_{x \to a} f(x)^{g(x)}$$

$$\ln L \quad = \quad \ln\left[\lim_{x \to a} f(x)^{g(x)} \right]$$

$$\ln L \quad = \quad \lim_{x \to a} \left[\ln\left(f(x)^{g(x)} \right) \right]$$

$$\ln L \quad = \quad \lim_{x \to a} \left[g(x) \cdot \ln f(x) \right]$$

$$\ln L \quad = \quad solution$$

$$e^{\ln L} \quad = \quad e^{(solution)}$$

$$L \quad = \quad e^{(solution)}$$

LH Rule: Indeterminate Exponents -- Ex. 6a

Find: $\displaystyle\lim_{x \to 0^+} (1 + \sin 3x)^{\cot x}$

Indeterminate exponential form.	$\displaystyle\lim_{x \to 0^+} (1 + \sin 3x)^{\cot x}$ $= (1+0)^\infty = 1^\infty$
Let y = term with exponent.	$y = (1 + \sin 3x)^{\cot x}$
Take the log of both sides.	$\ln y = \cot x \cdot \ln(1 + \sin 3x)$ $\ln y = \dfrac{\ln(1 + \sin 3x)}{\tan x}$
Take limit of both sides.	$\displaystyle\lim_{x \to 0^+} \ln y = \lim_{x \to 0^+} \dfrac{\ln(1 + \sin 3x)}{\tan x}$ $= \dfrac{\ln(1 + \sin 0)}{\tan 0} = \dfrac{0}{0}$
Indeterminate result. Apply LH rule. Continued ...	

LH Rule: Indeterminate Exponents -- Ex. 6b	
Previously found	$$\lim_{x \to 0^+} \ln y = \lim_{x \to 0^+} \frac{\ln(1 + \sin 3x)}{\tan x}$$ $$= \frac{\ln(1+0)}{0} = \frac{0}{0}$$
Apply LH Rule.	$$\lim_{x \to 0^+} \ln y = \lim_{x \to 0^+} \frac{(1 + \sin 3x)}{\tan x}$$ $$= \lim_{x \to 0^+} \frac{\frac{d}{dx}(\ln(1 + \sin 3x))}{\frac{d}{dx}(\tan x)}$$ $$= \lim_{x \to 0^+} \frac{\left[\frac{3\cos(3x)}{(1 + \sin(3x))}\right]}{\sec^2(x)}$$ $$= \lim_{x \to 0^+} \frac{\left[\frac{3}{1}\right]}{(1)} = 3$$
Use this log rule: $e^{\ln n} = n$	$$\lim_{x \to 0^+} \ln y = 3$$ $$\lim_{x \to 0^+} e^{\ln y} = e^3$$ $$\lim_{x \to 0^+} y = e^3$$

<u>Optimization Problems</u>

Optimization Problem -- Ex. 1a
Minimize Area of a Can

Minimize the cost of producing a cylindrical metal soda can. The can should hold 16 oz. of soda.

Use: $16\ oz. = 473.2\ mL$ and $1\ L = 1000\ cm^3$

Convert to volume to cm^3	$V = 0.4732\ L = 473.2\ cm^3$
Vol. is given so relate vol. to can dimensions.	$V = (base\ area) \cdot h$ $V = (\pi r^2) \cdot h$ $473.2 = (\pi r^2) \cdot h$
Total Area is what we want to minimize so write eqn. for it. $A = f(r)$	$A = 2(base\ area) + side$ $A = 2(\pi r^2) + (2\pi r)h$ $A = 2(\pi r^2) + (2\pi r)\left[\frac{473.2}{\pi r^2}\right]$ $A = 2\pi r^2 + \frac{946.4}{r}$
Find A'	$A = 2\pi r^2 + (946.4)r^{-1}$ $A' = 4\pi r - (946.4)r^{-2}$

Continued ...

Optimization Problem -- Ex. 1b
Minimize Area of a Can

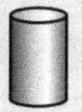

Set $A' = 0$ To find extrema.	$A = 2\pi r^2 + (946.4)r^{-1}$ $A' = 4\pi r - (946.4)r^{-2} = 0$
Solve for r	$r^{-2}(4\pi r^3 - 946.4) = 0$ $4\pi r^3 - 946.4 = 0$ $r = \sqrt[3]{\dfrac{946.4}{4\pi}} = 4.22\ cm$
Solve for h	$h = \dfrac{473.2}{\pi r^2}$ (found previously) $h = \dfrac{473.2}{\pi(4.22)^2} = 8.46\ cm$
To verify it is a minimum, check concavity at $r = 4.22$	$A' = 4\pi r - (946.4)r^{-2}$ $A'' = 4\pi + 2(946.4)r^{-3}$ $A''(4.22) = $ Positive Concave up $\rightarrow$ Minium

Optimization Problem -- Ex. 2a
Minimize Distance

Find the point on the line $y = 2x$
That is closest to the point $(10, 1)$

Distance is what we want to minimize so write an eqn. for it.	Let: (x, y) be a point on line. $d = \sqrt{(\Delta x)^2 + (\Delta y)^2}$ $d = \sqrt{(x - 10)^2 + (y - 1)^2}$ $d = \sqrt{x^2 + y^2 - 20x - 2y + 101}$
Substitute $y = 2x$ To get $d = f(x)$	$d = \sqrt{x^2 + 4x^2 - 20x - 4x + 101}$ $d = \sqrt{5x^2 - 24x + 101}$ $d = (5x^2 - 24x + 101)^{1/2}$
Find d'	$d' =$ $\frac{1}{2}(5x^2 - 24x + 101)^{\frac{-1}{2}}(10x - 24)$

Continued ...

Optimization Problem -- Ex. 2b
Minimize Distance

Previously Found	$d' =$ $\frac{1}{2}(5x^2 - 24x + 101)^{\frac{-1}{2}}(10x - 24)$
Extrema when $d' = 0$	$d' = \dfrac{(10x - 24)}{2\sqrt{5x^2 - 24x + 101}}$ $0 = 10x - 24$ $x = \dfrac{24}{10} = 2.4$
Find y	$y = 2x$ $y = 2(2.4) = 4.8$
Conclusion: Minimum distance between point and line ≈ 8.5	Point $(x, y) = (2.4, 4.8)$ Is the point on the line $y = 2x$ Closest to point $(10, 1)$ $d = \sqrt{(2.4 - 10)^2 + (4.8 - 1)^2}$ $d = \sqrt{(-7.6)^2 + (3.8)^2} \approx 8.5$

Newton's Method

Newton's Method -- Derivation

If $f(x) = 0$

Then $x_{n+1} = x_n - \dfrac{f(x_n)}{f'(x_n)}$

The general idea for how the equation was derived is shown below. (An abbreviated proof)

$(x_1, y_1) = 1^{st}$ Approx.	$f' = slope = \dfrac{\Delta y}{\Delta x}$
$(x_2, y_2) = 2^{nd}$ Approx.	$f' = \dfrac{y_2 - y_1}{x_2 - x_1}$

Suppose our next guess is perfect, so $y_2 = 0$	$x_2 - x_1 = \dfrac{y_2 - f(x_1)}{f'}$
	$x_2 = x_1 - \dfrac{f(x_1)}{f'(x_1)}$
Keep going until you find x_n and x_{n+1} changing very little.	$\ldots$
	$x_{n+1} = x_n - \dfrac{f(x_n)}{f'(x_n)}$

Newton's Method -- Ex. 1a

Use Newton's Method: $\quad x_{n+1} = x_n - \dfrac{f(x_n)}{f'(x_n)}$

Approximate $\sqrt[4]{3}$ $\qquad$ To 2 decimal places.

Let x = what we want to find. Then, rearrange.	$x = \sqrt[4]{3}$ $x^4 = 3$ $x^4 - 3 = 0$	Required Format: $f(x) = 0$
Let the equation $= f(x) = 0$	$f(x) = x^4 - 3$ $f'(x) = 4x^3$	
First guess is $x_1 = 1$ Find next guess	$x_{n+1} = x_n - \dfrac{f(x_n)}{f'(x_n)}$ $x_2 = 1 - \dfrac{f(1)}{f'(1)}$ $x_2 = 1 - \dfrac{-2}{4} = 1.5$	
$x_2 = 1.5$ Find next guess	Continued ...	

Newton's Method -- Ex. 1b	
Approximate $\sqrt[4]{3}$	To 2 decimal places.

Previously, we found.	$f(x) = x^4 - 3$ $f'(x) = 4x^3$
$x_2 = 1.5$ Find next guess	$x_3 = 1.5 - \dfrac{f(1.5)}{f'(1.5)}$ $x_3 = 1.34722$
$x_3 = 1.3472$ Find next guess	$x_4 = 1.3472 - \dfrac{f(1.3472)}{f'(1.3472)}$ $x_4 = 1.3194$
$x_4 = 1.3194$ Find next guess	$x_5 = 1.3194 - \dfrac{f(1.3194)}{f'(1.3194)}$ $x_5 = 1.3161$
We're done!	Successive guesses constant to 2 decimal places.
Solution, rounded to 2 decimal places.	$x \approx 1.32$ $\sqrt[4]{3} \approx 1.32$

Antiderivatives

Antiderivative Theorem

If : $F(x)$ is an antiderivative

of $f(x)$ on an interval I

Then: The most general antiderivative

of $f(x)$ on I is: $F(x) + C$

If you take the derivative of the most general antiderivative, you will get $f(x)$.

$$\frac{d}{dx} [F(x) + C] = f(x)$$

"C" is the Constant of Integration.

Antiderivatives are the opposite of derivatives.
Later, you will see that antiderivatives are integrals.

Antiderivatives are another name for Integrals.

- **Derivatives Decrease** the degree of a function.

- **Integrals Increase** the degree of a function.

Antiderivative Table

Function	Particular Antiderivative
$c \cdot f(x)$	$c \cdot F(x)$
$f(x) + g(x)$	$F(x) + G(x)$
x^n	$\dfrac{x^{n+1}}{n+1}$
$\dfrac{1}{x}$	$\ln(x)$
$\cos(x)$	$\sin(x)$
$\sin(x)$	$-\cos(x)$
$\sec^2(x)$	$\tan(x)$

Antiderivatives – Ex. 1	
Given: $f'(x) = e^x + 25x^3$ and $f(0) = -2$	
Find: $f(x)$	

Find the antiderivative.	$f'(x) = e^x + 25x^3$ $f(x) = e^x + 25\dfrac{x^4}{4} + C$
Use the given point $(0, -2)$ to solve for C (Constant of integration)	$f(0) = -2$ Given $e^0 + 25\dfrac{0^4}{4} + C = -2$ $1 + 0 + C = -2$ $C = -3$
General Solution	$f(x) = e^x + 25\dfrac{x^4}{4} + C$
Particular Solution	$f(x) = e^x + 25\dfrac{x^4}{4} - 3$

Antiderivatives – Ex. 2a	
Find: $f(x)$	Given: $f''(x) = 12x^2 + 6x - 4$
	And $f(0) = 9, \quad f(1) = 1$

Find the antideriv. of f'' to get f'	$f''(x) = 12x^2 + 6x - 4$ $f'(x) = 12\frac{x^3}{3} + 6\frac{x^2}{2} - 4x + C$ $f'(x) = 4x^3 + 3x^2 - 4x + C$
Find the antideriv. of f' to get f	$f = 4\frac{x^4}{4} + 3\frac{x^3}{3} - 4\frac{x^2}{2} + Cx + D$
Use given points to find C and D.	Continued ...

Antiderivatives – Ex. 2b	
Find: $f(x)$	Given: $f''(x) = 12x^2 + 6x - 4$ And $f(0) = 9, \quad f(1) = 1$

Previously Found	$f = 4\dfrac{x^4}{4} + 3\dfrac{x^3}{3} - 4\dfrac{x^2}{2} + Cx + D$
Use given points to find C and D.	$f(0) = 9$ Given $0 + 0 - 0 + 0 + D = 9 \quad \rightarrow \boldsymbol{D = 9}$
	$f(1) = 1$ Given $4\dfrac{1}{4} + 3\dfrac{1}{3} - 4\dfrac{1}{2} + C(1) + 9 = 1$ $1 + 1 - 2 + C + 9 = 1 \quad \rightarrow \boldsymbol{C = -8}$
Particular Solution	$f = 4\dfrac{x^4}{4} + 3\dfrac{x^3}{3} - 4\dfrac{x^2}{2} - 8x + 9$

Integrals

<u>Area</u>

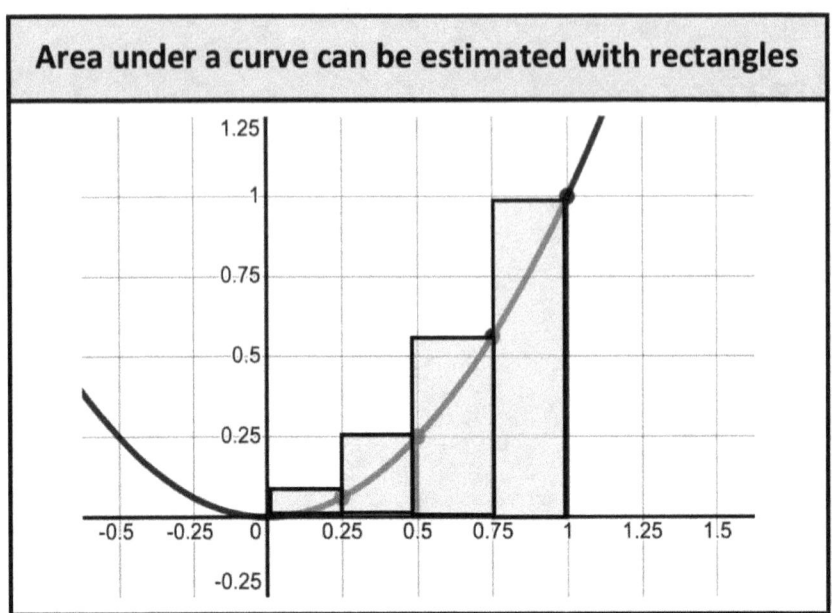

Area under a curve can be estimated with rectangles

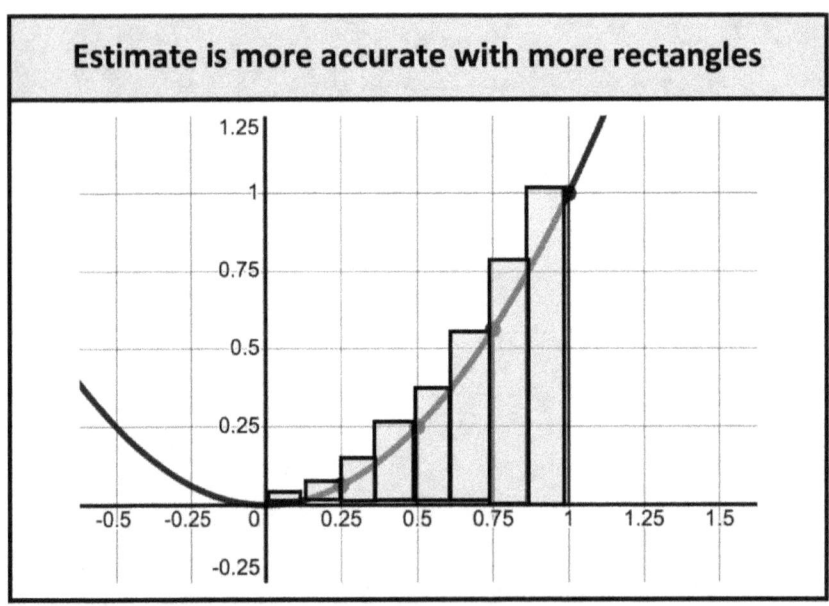

Estimate is more accurate with more rectangles

Riemann Sum
Used to Estimate the Area Under a Curve

Riemann Sum $= \sum_{i=1}^{n} (height)_i \cdot (width)_i$

$$= \sum_{i=1}^{n} f(x_i{}^*) \, \Delta x$$

Notes:

- The width (Δx) is constant if interval is evenly divided into n intervals.

- The height could be based on the right, left, or mid-points of the intervals. That's why there is an asterisk in the Riemann Sum equation.

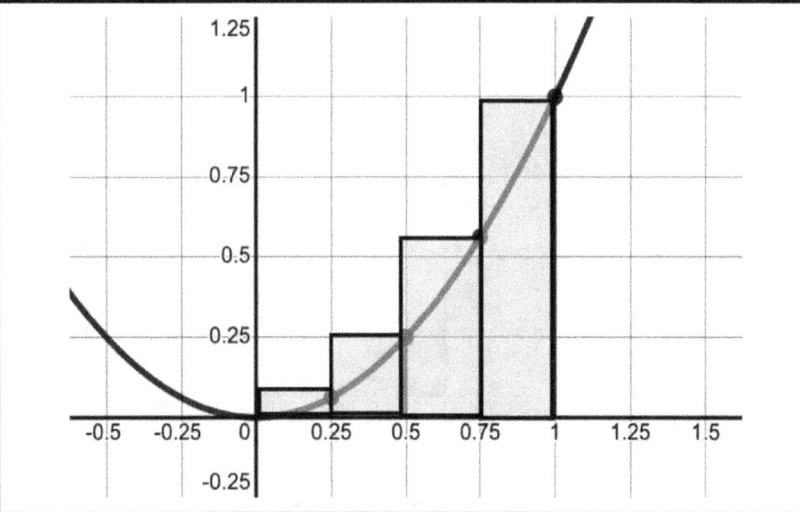

Estimate Area Under a Curve – Ex. 1a

$f(x) = x^2$
on $[1, 1]$

An Integral
can calculate
the area under
the curve.

$A = \frac{1}{3} = 0.\overline{3}$

To estimate
the area,
first break
the interval
into n
equal
sections.

Here, $n = 4$

Width $= \frac{1}{4}$

Estimate Area Under a Curve – Ex. 1b
With Right-End Points

$f(x) = x^2$
on $[1,1]$
$n = 4$

Rectangles
with
**Right End
Points**

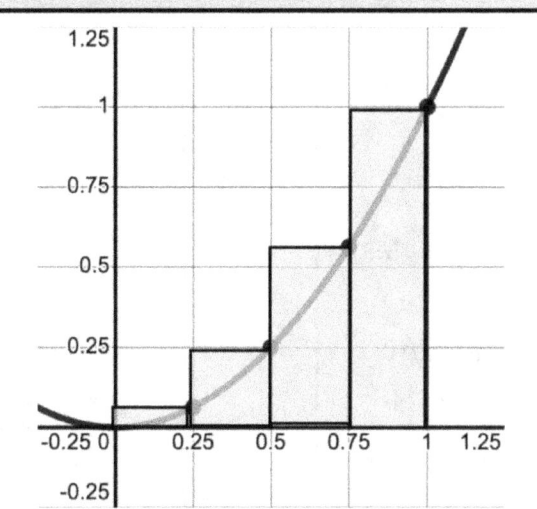

Find
area of
rectangles
to estimate
area.

width $= \frac{1}{4}$

$A = \sum_{i=1}^{4} (l_i)(w_i)$

$A = w \sum_{i=1}^{4} (l_i)$

$A = \frac{1}{4} \sum_{i=1}^{4} f(x_i)$

$A = \frac{1}{4} \left[\left(\frac{1}{4}\right)^2 + \left(\frac{2}{4}\right)^2 + \right.$

$\left. + \left(\frac{3}{4}\right)^2 + \left(\frac{4}{4}\right)^2 \right]$

$= .25 \,(1.875) = 0.4688$

(Estimate is more than actual.)

Estimate Area Under a Curve – Ex. 1c
With Left-End Points

$f(x) = x^2$
on $[1,1]$
$n = 4$

Rectangles
with
Left End
Points

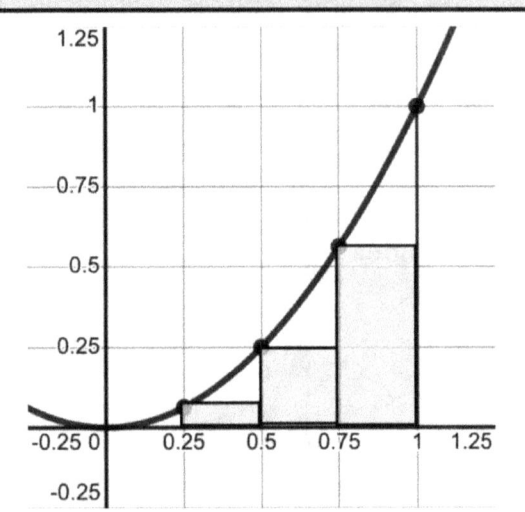

Find
area of
rectangles
to estimate
area.

width $= \frac{1}{4}$

$A = \sum_{i=1}^{4} (l_i)(w_i)$

$A = w \sum_{i=1}^{4} (l_i)$

$A = \frac{1}{4} \sum_{i=1}^{4} f(x_i)$

$A = \frac{1}{4} \left[(0)^2 + \left(\frac{1}{4}\right)^2 + \right.$

$\left. + \left(\frac{2}{4}\right)^2 + \left(\frac{3}{4}\right)^2 \right]$

$A = .25(0.875) = 0.2188$

(Estimate is less than actual.)

Estimate Area Under a Curve – Ex. 1d **With Mid-Points**	
$f(x) = x^2$ on $[1,1]$ $n = 4$ Rectangles with **Mid-Points**	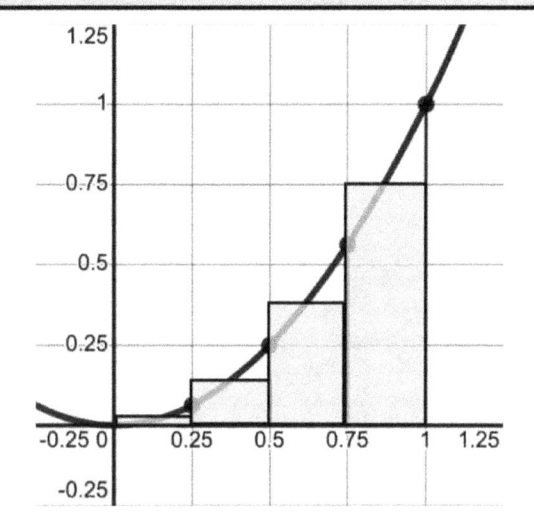
Find area of rectangles to estimate area. width $= \frac{1}{4}$	$A = \sum_{i=1}^{4} (l_i)(w_i)$ $A = w \sum_{i=1}^{4} (l_i)$ $A = \frac{1}{4} \sum_{i=1}^{4} f(x_i)$ $A = \frac{1}{4} \left[\left(\frac{1}{8}\right)^2 + \left(\frac{3}{8}\right)^2 + \right.$ $\left. + \left(\frac{5}{8}\right)^2 + \left(\frac{7}{8}\right)^2 \right]$ $A = .25(1.3125) = 0.3281$ (Estimate is closer to actual.)

Definite Integrals (Riemann Sum)

Definition of Definite Integral (Riemann Sum)

$$\int_a^b f(x)\ dx \ = \ \lim_{n \to \infty} \ \sum_{i=1}^{\infty} \ f(x_i^*)\ \Delta x$$

Where: $\Delta x \ = \ \dfrac{b-a}{n}$

In other words ...

If the area under a curve is broken into an infinite number of rectangles, then the actual area can be calculated.

Notes:

- It is a definite integral because the boundaries are specified. $[a, b]$
- In the summation, the asterisk $*$ indicates that each x may represent the left boundary, right boundary, or midpoint of the rectangular section.

Fundamental Theorem of Calculus

The Fundamental Theorem of Calculus -- Part 1

$$g(x) = \int_a^x f(t)\, dt \qquad a \le x \le b$$

If $\qquad f$ is continuous on $[a, b]$

Then

$\qquad g$ is continuous on $[a, b]$ and

$\qquad g$ is differentiable on (a, b)

$\qquad$ And $\quad g'(x) = f(x)$

In other words ...

$\qquad$ Integrals are associated with derivatives.

The Fundamental Theorem of Calculus -- Part 2

$$\int_a^b f(x)\, dx = F(b) - F(a)$$

Where: $\quad F$ is any antiderivative of f

(i.e.) $\qquad f$ is a function such that $F' = f$

In other words ...

$\qquad$ The definite integral is the difference of the
$\qquad$ antiderivatives, evaluated at the boundaries.

The Fundamental Theorem of Calculus -- Ex. 1
Using FTC Part 1

Find the derivative of: $g(s) = \int_5^s (t - t^2)^8 \, dt$

| Solution | $g'(s) = (s - s^2)^8$ |

Find the derivative of: $g(x) = \int_0^x \sqrt{t + t^5} \, dt$

| Solution | $g'(x) = \sqrt{x + x^5}$ |

Find the derivative of: $h(x) = \int_1^{e^x} \ln t \, dt$

| Must use the chain rule | Let $u = e^x$ $\dfrac{du}{dx} = e^x$ |

$h'(x) = \dfrac{d}{dx}\left[\int_1^{e^x} \ln t \, dt \right]$

$= \dfrac{d}{du}\left[\int_1^u \ln t \, dt \right]\dfrac{du}{dx}$

$= [\ln u]\dfrac{du}{dx} = [\ln(e^x)](e^x) = x\, e^x$

Indefinite Integrals

Indefinite Integral

$$\int f(x)\, dx \;=\; F(x) \qquad \leftrightarrow \qquad F'(x) = f(x)$$

Indefinite Integral

In other words ...

An <u>indefinite integral</u> is a <u>function</u>.

A <u>definite integral</u> is a <u>number</u>.

Let $F(x) =$ indefinite integral function , then

The definite integral is a number $= F(b) - F(a)$

Some Examples of Indefinite Integrals

$\int x^2\, dx \;=\; \dfrac{x^3}{3} + C$	Because $\dfrac{d}{dx}\left(\dfrac{x^3}{3} + C\right) = x^2$
$\int 5x\, dx \;=\; 5\dfrac{x^2}{2} + C$	Because $\dfrac{d}{dx}\left(5\dfrac{x^2}{2} + C\right) = 5x$

Integration – Ex. 1		
Use the "Power Rule" To find the Indefinite Integral for the integral.		

Integral	Power Rule	$\int x^n \, dx = \dfrac{x^{n+1}}{n+1} + C$
$\int x^3 \, dx$	$= \dfrac{x^4}{4} + C$	
$\int 5x^3 \, dx$	$= 5 \cdot \dfrac{x^4}{4} = \dfrac{5x^4}{4} + C$	
$\int x^{-3} \, dx$	$= \dfrac{x^{-2}}{-2} = -\dfrac{1}{2x^2} + C$	
$\int \dfrac{5}{x^3} \, dx$	$= \int 5x^{-3} \, dx = 5 \cdot \dfrac{x^{-2}}{-2}$ $= -\dfrac{5}{2x^2} + C$	
$\int \sqrt{x} \, dx$	$= \int x^{\frac{1}{2}} \, dx = \dfrac{x^{\frac{3}{2}}}{\left(\frac{3}{2}\right)}$ $= \left(\dfrac{2}{3}\right) x^{\frac{3}{2}} + C$	

Integration – Particular Solution -- Ex. 2

Given: $F'(x) = \frac{1}{x^2}$, $x > 0$, $F(1) = 0$

Find: The general and particular solutions.

$F(x) = \int F'(x)\, dx$

$F(x) = \int \left[\frac{1}{x^2} \right] dx$

$F(x) = \int \left[x^{-2} \right] dx$

$F(x) = \frac{x^{-1}}{-1} + C$

$F(x) = -\frac{1}{x} + C$ **General Solution**

Use Given Initial Condition: $(x, y) = (1, 0)$

$0 = -\frac{1}{1} + C \quad \rightarrow \quad C = 1$

$F(x) = -\frac{1}{x} + 1$ **Particular Solution**

Integration Tables

Integration -- Properties
$u = f(x)$ and $v = v(x)$

$\int_a^b k\,u\,dx$	$=$	$k \int_a^b u\,dx$
$\int_a^b [u \pm v]\,dx$	$=$	$\int_a^b u\,dx \pm \int_a^b v\,dx$
$\int_a^b c\,dx$	$=$	$c(b - a)$
$\int_a^b u\,dx$	$=$	$-\int_b^a u\,dx$

Integration Table – Exponents & Logs	
$\int du$	$=\quad u + C$
$\int u^n \, du$	$=\quad \dfrac{u^{n+1}}{n+1} + C$
$\int \dfrac{1}{u} \, du$	$=\quad \ln|u| + C$
$\int e^u \, du$	$=\quad e^u + C$
$\int a^u \, du$	$=\quad \left(\dfrac{1}{\ln a}\right) a^u + C$

Integration Table – Trig Functions		
$\int \cos u \; du$	$=$	$\sin u + C$
$\int \sin u \; du$	$=$	$-\cos u + C$
$\int \cot u \; du$	$=$	$\ln\lvert \sin u \rvert + C$
$\int \tan u \; du$	$=$	$-\ln\lvert \cos u \rvert + C$
$\int \sec u \; du$	$=$	$\ln\lvert \sec u + \cot u \rvert + C$
$\int \csc u \; du$	$=$	$-\ln\lvert \csc u + \cot u \rvert + C$
$\int \sec^2 u \; du$	$=$	$\tan u + C$
$\int \csc^2 u \; du$	$=$	$-\cot u + C$
$\int \sec u \tan u \; du$	$=$	$\sec u + C$
$\int \csc u \cot u \; du$	$=$	$-\csc u + C$

Integration Table – Inverse Trig Functions	
$\int \dfrac{1}{\sqrt{a^2 - u^2}}\, du$	$= \quad \arcsin\left(\dfrac{u}{a}\right) + C$
$\int \dfrac{1}{a^2 + u^2}\, du$	$= \quad \left(\dfrac{1}{a}\right) \arctan\left(\dfrac{u}{a}\right) + C$
$\int \dfrac{1}{u\sqrt{u^2 - a^2}}\, du$	$= \quad \arcsin\left(\dfrac{u}{a}\right) + C$

Integration Table – Extra Functions	
$\int \ln x\ dx$	$= \quad \tan x + C$
$\int \sec x\ dx$	$= \quad \ln\lvert \sec x + \cot x \rvert + C$
$\int \sec^2 x\ dx$	$= \quad \tan x + C$
$\int \sec^3 x\ dx$	
$= \dfrac{1}{2}\,[\sec x \cdot \tan x$	
$\qquad + \ln\lvert \sec x + \tan x \rvert\,]\ + C$	

Substitution Rule

Integration – Substitution Rule

$$\int f\big(g(x)\big) \cdot g'(x) \;=\; \int f(u) \cdot du$$

When integrating, make sure it is In the
correct form: $\int f(u) \cdot du$

Definite Integrals – Substitution Rule

$$\int_a^b f\big(g(x)\big) \cdot g'(x)\, dx \;=\; \int_{u_1}^{u_2} f(u) \cdot du$$

Where: $u_1 = g(a)$

and $\quad u_2 = g(b)$

When using substitution,
recalculate the boundaries to be in terms of u .

	Integration – Substitution – Ex. 1
	Find: $I = \int x^3 (x^4 + 2)^5 \, dx$

Identify u	$u = x^4 + 2$
Find du	$\dfrac{du}{dx} = 4x^3$ $du = 4x^3 \, dx$
Rewrite integral in terms of u & du	$I = \int u^5 \, (x^3 dx)$ $I = \frac{1}{4} \cdot \int u^5 \, (4x^3 dx)$ $I = \frac{1}{4} \cdot \int u^5 \, du$
Evaluate integral.	$I = \frac{1}{4} \cdot \int u^5 \, du = \frac{1}{4} \left[\dfrac{u^6}{6} \right] + C$
Substitute for u	$I = \frac{1}{24} (x^4 + 2)^6 + C$

Integration – Substitution – Ex. 2
Find: $\quad I = \int e^{5x+2}\, dx$

Identify u	$u = 5x + 2$
Find du	$\dfrac{du}{dx} = 5$ $du = 5\, dx$
Rewrite integral in terms of u & du	$I = \int e^u\,(dx)$ $I = \frac{1}{5} \cdot \int e^u\,(5\, dx)$ $I = \frac{1}{5} \cdot \int e^u\, du$
Evaluate integral.	$I = \frac{1}{5} \cdot \int e^u\, du = \frac{1}{5}\,[e^u] + C$
Substitute for u	$I = \frac{1}{5}\,[e^{5x+2}] + C$

Integration – Substitution – Ex. 3
Find: $I = \int \cos^3 x \; dx$

Identify u	$\int \cos x \cdot \cos^2 x \; dx$ $\int \cos x \cdot (1 - \sin^2 x) \; dx$ $\int \cos x \; dx - \int \sin^2 x \cdot \cos x \; dx$ $u = \sin x$
Find du	$\dfrac{du}{dx} = \cos x$ $du = \cos x \; dx$
Rewrite integral and solve	$I = \int \cos x \; dx - \int u^2 \; du$ $I = \sin x - \dfrac{u^3}{3} + C$
Substitute for u	$I = \sin x - \dfrac{\sin^3 x}{3} + C$

<u>Negative and Positive Integrals</u>

Positive and Negative Integrals

Positive Integrals	Negative Integrals
$f(x) \geq 0$ on $[a,b]$	$f(x) \leq 0$ on $[a,b]$
$\int_a^b f(x)\,dx \geq 0$	$\int_a^b f(x)\,dx \leq 0$

Integration – Symmetry Rules

Even $f(x)$	Odd $f(x)$
$f(-x) = f(x)$	$f(-x) = -f(x)$
$\int_{-a}^{a} f(x)\, dx$ $= 2\int_{0}^{a} f(x)\, dx$	$\int_{-a}^{a} f(x)\, dx = 0$

Integration – $\int \sin x \, dx$	
$f(x) = \sin x$	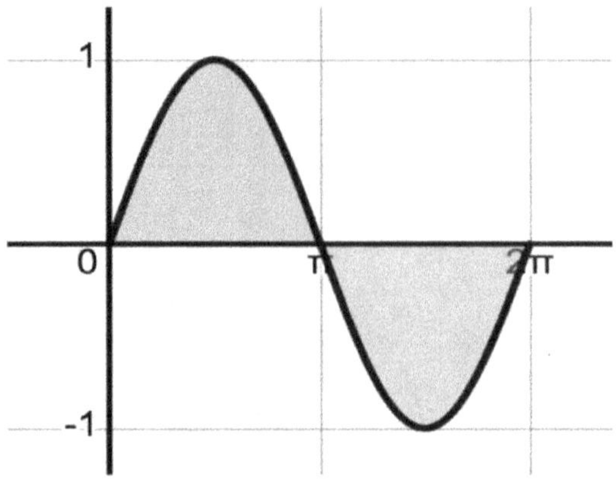
Positive and Negative Areas	Areas above the x-axis are often called positive areas. Areas below the x-axis are often called negative areas.
Area under the curve.	Integrals are often described as the area under a curve.
Think about this!	The area under a curve is infinite. Where does it end?
Actually, Integrals are the area BETWEEN two curves.	The above graph shows the area between two curves: $y_1 = \sin x$ and $y_2 = 0$

<u>Definite Integrals (Apply FTC 2)</u>

Definite Integral of f from a to b

$$\int_a^b f(x)\, dx \;=\; \lim_{n \to \infty} \sum_{i=1}^{n} f(x_i^*)\, \Delta x$$

$n \;=\;$ Number of Intervals

$$\Delta x \;=\; \frac{b-a}{n} \;=\; \text{Width of rectangle}$$

$f(x_i^*) \;=\;$ Height of rectangle

Recall:
The Fundamental Theorem of Calculus Part 2

$$\int_a^b f(x)\, dx \;=\; F(b) - F(a)$$

Where F is any antiderivative of f.

(i.e.) f is a function such that $F' = f$

In other words ... The <u>definite integral</u> is the difference of the antiderivatives, evaluated at the boundaries.

Definite Integral -- Ex. 1	
Find: $\int_0^2 x^3\, dx$	

Let "I" represent the original integral. This is just a style choice!	$I = \int_0^2 x^3\, dx$
Find the integral (antiderivative) for the function.	$I = \int_0^2 x^3\, dx$ $I = \left[\frac{x^4}{4} \right]_0^2 = \frac{1}{4}[\, x^4 \,]_0^2$
Apply FTC part 2 $\int_a^b f(x)\, dx$ $= F(b) - F(a)$	$I = \frac{1}{4}[\, 2^4 - 0^4 \,]$ $I = \frac{1}{4}[\, 16 - 0 \,]$ $I = \frac{1}{4}(16) = 4$

Definite Integral -- Ex. 2

$$\text{Find: } \int_1^2 \frac{5}{x^3}\, dx$$

Rewrite function so It is easier to integrate.	$$I = \int_1^2 5x^{-3}\, dx$$
Integrate.	$$I = \left[5\, \frac{x^{-2}}{-2}\right]_1^2 = \frac{5}{-2}\left[\frac{1}{x^2}\right]_1^2$$
Apply FTC part 2 $$\int_a^b f(x)\, dx$$ $$= F(b) - F(a)$$	$$I = -\frac{5}{2}\left[\frac{1}{2^2} - \frac{1}{1^2}\right]$$ $$I = -\frac{5}{2}\left[\frac{1}{4} - \frac{4}{4}\right]$$ $$I = -\frac{5}{2}\left[-\frac{3}{4}\right] = \frac{15}{8}$$

Definite Integral -- Ex. 3a	
Find: $\int_1^e \frac{\ln x}{x}\, dx$	

Use u-sub Identify u and du	$u = \ln x$ $\frac{du}{dx} = \frac{1}{u} \quad \rightarrow \quad du = \frac{1}{u}\, dx$
Rewrite in terms of u and calculate new boundaries. Remember: $u = \ln x$	$I = \int_{x=1}^{x=e} \ln x \left(\frac{1}{x}\, dx \right)$ $I = \int_{u=\ln 1}^{u=\ln e} u\, du$ $I = \int_{u=0}^{u=1} u\, du$ $I = \int_0^1 u\, du$
Apply FTC part 2 $\int_a^b f(x)\, dx$ $= F(b) - F(a)$	$I = \left[\frac{u^2}{2} \right]_0^1$ $I = \frac{1}{2}[\, 1^2 - 0^2\,] = \frac{1}{2}$

Definite Integral -- Ex. 3b. (Slightly Different Sol'n)

Find: $\int_1^e \frac{\ln x}{x} dx$

Use u-sub Identify u and du	$u = \ln x$ $\frac{du}{dx} = \frac{1}{u} \quad \rightarrow \quad du = \frac{1}{u} dx$
Rewrite in terms of u. No boundaries.	$I = \int_{x=1}^{x=e} \ln x \left(\frac{1}{x} dx \right)$ $I = \int u \, du$
Apply FTC 2 Return the x-boundaries $\int_a^b f(x) dx$ $= F(b) - F(a)$	$I = \left[\frac{u^2}{2} \right]$ $I = \left[\frac{(\ln x)^2}{2} \right]_{x=1}^{x=e}$ $I = \frac{1}{2} \left[(\ln e)^2 - (\ln 1)^2 \right]$ $I = \frac{1}{2} \left[1^2 - 0^2 \right] = \frac{1}{2}$

Definite Integral -- Ex. 4a

Find the area between
the two curves:

$y_1 = \sin x$ and $y_2 = 0$

For: $0 \le x \le 2\pi$

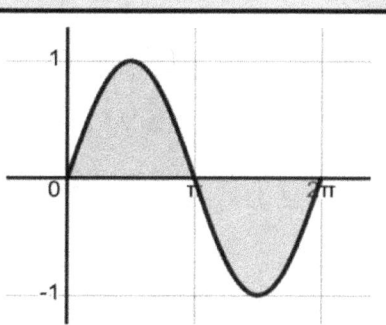

Calculate each half separately.

$A_1 = \int_0^\pi (y_1 - y_2)\,dx$	$A_2 = \int_\pi^{2\pi} (y_2 - y_1)\,dx$
$A_1 = \int_0^\pi (\sin x - 0)\,dx$	$A_2 = \int_\pi^{2\pi} (0 - \sin x)\,dx$
$A_1 = \int_0^\pi (\sin x)\,dx$	$A_2 = \int_\pi^{2\pi} (-\sin x)\,dx$
$A_1 = [-\cos x\,]_0^\pi$	$A_2 = [\cos x\,]_\pi^{2\pi}$
$A_1 = -[\cos \pi - \cos 0\,]$	$A_2 = [\cos 2\pi - \cos \pi\,]$
$A_1 = -[(-1) - (1)\,]$	$A_2 = [(1) - (-1)\,]$
$A_1 = -[-2\,] = 2$	$A_2 = 2$

Total Area between the two curves.	$A = A_1 + A_2$ $A = 2 + 2$ $A = 4$

Definite Integral -- Ex. 4b (Slightly Different Sol'n)

Find the area between
the two curves:

$y_1 = \sin x$ and $y_2 = 0$

For: $0 \leq x \leq 2\pi$

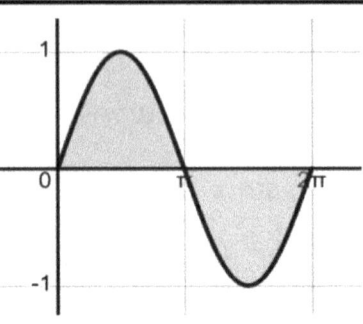

Use Symmetry.

$$A = 2 \int_0^{\pi} (y_1 - y_2)\, dx$$

$$A = 2 \int_0^{\pi} (\sin x - 0)\, dx$$

$$A = 2 \int_0^{\pi} (\sin x)\, dx$$

$$A = 2 \left[-\cos x \right]_0^{\pi}$$

$$A = -2 \left[\cos \pi - \cos 0 \right]$$

$$A = -2 \left[(-1) - (1) \right]$$

$$A = -2 \left[-2 \right] \; = \; 4$$

Definite Integral -- Ex. 4c   (WRONG Solution)

Find the area between
the two curves:

$y_1 = \sin x$ and $y_2 = 0$

For: $0 \leq x \leq 2\pi$

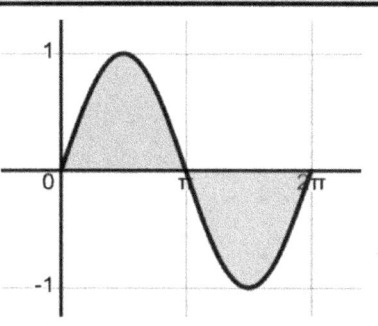

WRONG Solution and WRONG Answer!!!

$A = \int_0^{2\pi}(y_1 - y_2)\,dx = \int_0^{2\pi}(\sin x - 0)\,dx$

$A = \int_0^{2\pi}(\sin x)\,dx = [\cos x]_0^{2\pi}$

$A = [\cos 2\pi - \cos 0] = [(1) - (1)] = 0$

$A = 0$ Wrong approach and wrong answer!

See previous page for correct solution & answer.

Correct answer: $A = 4$

Integration by Parts (IBP)

Integration By Parts (IBP) -- Equation
$$\int u \, dv \; = \; uv - \int v \, du$$ If $u = f(x)$ and $v = g(x)$
In other words ... If you find it difficult to integrate, scrambling it up may help!

	Integration By Parts (IBP) Use "LIATE" to Pick the Parts
	$$\int u\,dv \;=\; uv - \int v\,du$$
Question	What is $u(x)$ and what is $v(x)$?
Answer	The first function listed in the acronym "LIATE" is the u part. The rest is the dv

L	=	Log Functions
I	=	Inverse Trig Functions
A	=	Algebraic Functions
T	=	Trig Functions
E	=	Exponential Functions

Integration By Parts (IBP) -- Ex. 1
Evaluate: $\quad I = \int \ln x \, dx$

Use LIATE to find u	$u = \ln x$ $\dfrac{du}{dx} = \dfrac{1}{x}$ $du = \dfrac{1}{x} \, dx$	$dv = dx$ $\int dv = \int dx$ $v = x$

IBP: $\quad \int u \, dv \;=\; uv - \int v \, du$

$I \;=\; \int \ln x \, dx \;=\; x \ln x \;-\; \int x \left(\dfrac{1}{x} dx \right)$

$I \;=\; x \ln x \;-\; \int (1) \, dx$

$I \;=\; x \ln x \;-\; x \;+\; C$

Helpful Notes:
- Let I or I_1 represent original integral.
- You may need to apply IBP several times.
- Name additional integrals I_2, I_3, ...

Integration By Parts (IBP) -- Ex. 2a

Evaluate: $I_1 = \int t^2 e^t \, dt$

Use LIA̱TE to find u	$u = t^2$ $\dfrac{du}{dt} = 2t$ $du = 2t \, dt$	$dv = e^t dt$ $\int dv = \int e^t dt$ $v = e^t$

IBP: $\int u \, dv = uv - \int v \, du$

$I_1 = \int t^2 e^t \, dt = t^2 e^t - \int e^t \, 2t \, dt$

$I_1 = t^2 e^t - \int e^t \, 2t \, dt$

$I_1 = t^2 e^t - 2 \int e^t t \, dt$

$I_1 = t^2 e^t - 2 I_2$

Use IBP again to solve: $I_2 = \int e^t t \, dt$

Continued ...

Integration By Parts (IBP) -- Ex. 2b

Evaluate: $I_2 = \int e^t t \, dt$

Use LIA**T**E to find u	$u = t$ $\frac{du}{dt} = 1$ $du = dt$	$dv = e^t \, dt$ $\int dv = \int e^t \, dt$ $v = e^t$

IBP: $\int u \, dv = uv - \int v \, du$

$I_2 = \int t e^t \, dt = te^t - \int e^t \, dt$

$I_2 = te^t - e^t$

$I_1 =$ Original Integral

$I_1 = t^2 e^t - 2 \cdot I_2$

$I_1 = t^2 e^t - 2[\, te^t - e^t \,]$

$I_1 = t^2 e^t - 2te^t + 2e^t \quad + \quad C$

Integration By Parts (IBP) -- Ex. 3

Evaluate: $I = \int x \sin x \, dx$

Use LIATE to find u	$u = x$ $\dfrac{du}{dx} = 1$ $du = dx$	$dv = \sin x \, dx$ $\int dv = \int \sin x \, dx$ $v = -\cos x$

IBP: $\int u \, dv = uv - \int v \, du$

$I = x(-\cos x) - \int(-\cos x) \, dx$

$I = -x \cos x + \sin x$

$I = -x \cos x + \sin x + C$

Integration By Parts (IBP) -- Ex. 4

Evaluate the Integral: $\int x^3 \ln x \; dx$

Use LIATE to find u	$u = \ln x$ $du = \frac{1}{x} dx$	$dv = x^3 \; dx$ $\int dv = \int x^3 \; dx$ $v = \frac{x^4}{4}$

$I \;=\; uv - \int v \; du$

$I \;=\; (\ln x)\left(\frac{x^4}{4}\right) \;-\; \int \left(\frac{x^4}{4}\right)\left(\frac{1}{x} dx\right)$

$I \;=\; \frac{1}{4} x^4 \ln x \;-\; \frac{1}{4} \int x^3 \; dx$

$I \;=\; \frac{1}{4} x^4 \ln x \;-\; \frac{1}{4} \cdot \frac{x^4}{4} \;+\; C$

$I \;=\; \frac{1}{4} x^4 \ln x \;-\; \frac{1}{16} x^4 \;+\; C$

Integration By Parts (IBP) -- Ex. 5

Evaluate the Integral: $\int x^3 \cos x \; dx$

Use: LI **A** T E	$u = x^3$	$dv = \cos x \; dx$
to find **u**	$du = 3x^2 \; dx$	$\int dv = \int \cos x \; dx$
		$v = \sin x$

$I \; = \; uv - \int v \; du$

$I \; = \; x^3 \sin x \; - \; \int \sin x \cdot 3x^2 \; dx$

Use: LI **A** T E	$u = 3x^2$	$dv = \sin x \; dx$
to find **u**	$du = 6x \; dx$	$v = - \cos x$

$I \; = \; x^3 \sin x \; + \; 3x^2 \cos x \; - \; \int 6x \cos x \; dx$

Use: LI **A** T E	$u = 6x$	$dv = \cos x \; dx$
to find **u**	$du = 6 \; dx$	$v = \sin x$

$I = x^3 \sin x + 3x^2 \cos x - 6x \sin x \; + \int 6 \sin x \; dx$

$I = x^3 \sin x + 3x^2 \cos x - \; 6x \sin x - \; 6 \cos x \; + \; C$

$I = \sin x \, (x^3 - 6x) \; + \; \cos x \, (3x^2 - 6) \; + \; C$

Integration By Parts (IBP) -- Ex. 6

Evaluate the Integral: $\int x^2 \arctan x \, dx$

Use: L I A T E to find **u**	$u = \arctan x$ $du = \dfrac{1}{1+x^2} \, dx$	$dv = x^2 \, dx$ $\int dv = \int x^2 \, dx$ $v = \dfrac{1}{3} x^3$

$$I = \arctan x \cdot \frac{1}{3} x^3 \; - \; \int \frac{1}{3} x^3 \cdot \frac{1}{1+x^2} \, dx$$

$$I = \frac{1}{3} x^3 \arctan x \; - \; \frac{1}{3} \int \frac{x^3}{1+x^2} \, dx$$

Long Division	$\begin{array}{r} x \\ x^2 + 0x + 1 \,\overline{\big)\, x^3 + 0x^2 + 0x + 0} \\ -\ \underline{(x^3 + 0x\ \ + x)} \\ -x \end{array}$

$$I = \frac{1}{3} x^3 \arctan x \; - \; \frac{1}{3} \int \left(x - \frac{x}{x^2+1} \right) dx$$

$$I = \frac{1}{3} x^3 \arctan x \; - \; \frac{1}{3} \int x \, dx \; + \; \frac{1}{3} \int \left(\frac{x}{x^2+1} \right) dx$$

$$I = \frac{1}{3} x^3 \arctan x \; - \; \frac{1}{3} \cdot \frac{1}{2} x^2 \; + \; \frac{1}{3} \cdot \frac{1}{2} \int \left(\frac{1}{x^2+1} \right) 2x \, dx$$

$$I = \frac{1}{3} x^3 \arctan x \; - \; \frac{1}{6} x^2 \; + \; \frac{1}{6} \ln(x^2+1) \; + \; C$$

Integration By Parts (IBP) -- Ex. 7

Evaluate the Integral: $\int e^{3x} \sin 2x \ dx$

Use: L I A __T__ E to find u	$u = \sin 2x$ $du = \cos 2x \cdot 2 \ dx$	$dv = e^{3x} \ dx$ $v = \frac{1}{3} e^{3x}$

$I = \int e^{3x} \sin 2x \ dx$

$I = \sin 2x \cdot \frac{1}{3} e^{3x} - \int \frac{1}{3} e^{3x} \cdot \cos 2x \cdot 2 \ dx$

$I = \frac{1}{3} \sin 2x \cdot e^{3x} - \frac{2}{3} \int e^{3x} \cdot \cos 2x \ dx$

Use: L I A __T__ E to find u	$u = \cos 2x$ $du = - \sin 2x \cdot 2 \ dx$	$dv = e^{3x} \ dx$ $v = \frac{1}{3} e^{3x}$

$I = \frac{1}{3} \sin 2x \ e^{3x} - \frac{2}{3} \left[\cos 2x \ \frac{1}{3} e^{3x} + \int \frac{1}{3} e^{3x} \sin 2x \cdot 2 \ dx \right]$

$I = \frac{1}{3} e^{3x} \sin 2x - \frac{2}{9} e^{3x} \cos 2x - \frac{4}{9} \int e^{3x} \sin 2x \ dx$

$I = \frac{1}{3} e^{3x} \sin 2x - \frac{2}{9} e^{3x} \cos 2x - \frac{4}{9} I$

$\frac{13}{9} I = \frac{1}{3} e^{3x} \sin 2x - \frac{2}{9} e^{3x} \cos 2x$

$I = \frac{1}{13} e^{3x} [\ 3 \sin 2x - 2 \cos 2x \ ] + C$

Integration By Parts (IBP) -- Ex. 8

Evaluate the Integral: $\int x \cos x \, dx$

Use: L I **A** T E to find **u**	$u = x$ $du = 1 \, dx$	$dv = \cos x \, dx$ $\int dv = \int \cos x \, dx$ $v = \sin x$

$I = uv - \int v \, du$

$I = (x)(\sin x) - \int(\sin x)(1 \, dx)$

$I = x \sin x + \int \sin x \, dx$

$I = x \sin x + \cos x + C$

Integration By Parts (IBP) -- Ex. 9

Evaluate the Integral: $\int \arctan x \; dx$

| Use: L I̲ A T E to find u | $u = \arctan x$ $du = \dfrac{1}{1+x^2} \; dx$ | $dv = (1) \; dx$ $\int dv = \int 1 \; dx$ $v = x$ |

$I = uv - \int v \; du$

$I = (\arctan x)(x) - \int (x)\left(\dfrac{1}{1+x^2} \; dx\right)$

$I = x \arctan x - \dfrac{1}{2} \int \dfrac{1}{1+x^2} \; 2x \; dx$

$I = x \arctan x - \dfrac{1}{2} \ln| \, 1 + x^2 \, | \; + \; C$

$I = x \arctan x - \ln \sqrt{1 + x^2} \; + \; C$

Integration By Parts (IBP) -- Ex. 10

Evaluate the Integral: $\int x \sin 2x \, dx$

| Use: L I **A** T E to find **u** The rest is dv | $u = x$ $du = dx$ | $dv = \sin 2x \, dx$ $\int dv = \int 2 \sin x \cos x \, dx$ $\int dv = 2 \int (\sin x) \cos x \, dx$ $\int dv = 2 \int (w) \, dw$ $2 \left(\frac{w^2}{2} \right) = w^2$ $v = \sin^2 x$ |

$I = uv - \int v \, du$

$I = x \sin^2 x - \int \sin^2 x \, dx$

$I = x \sin^2 x - \int \left(\frac{1 - \cos 2x}{2} \right) dx$

$I = x \sin^2 x - \frac{1}{2} \int 1 \, dx + \frac{1}{2} \int \cos 2x \, dx$

$I = x \sin^2 x - \frac{1}{2} x + \frac{1}{4} \int \cos 2x \, 2 \, dx$

$I = x \left(\frac{1 - \cos 2x}{2} \right) - \frac{1}{2} x + \frac{1}{4} \sin 2x$

$I = -\frac{1}{2} \cos 2x + \frac{1}{4} \sin 2x + C$

Integration By Parts (IBP) -- Ex. 11

Evaluate the Integral: $\int x e^{4x}\, dx$

| Use: L I $\underline{\textbf{A}}$ T E
to find $\boldsymbol{u}$

The rest is
dv | $u = x$

$du = 1\, dx$ | $dv = e^{4x}\, dx$
$\int dv = \int e^{4x}\, dx$
$\int dv = \frac{1}{4} \int e^{4x}\, 4\, dx$
$v = \frac{1}{4} e^{4x}$ |

$I \;=\; uv - \int v\, du$

$I \;=\; x\, \frac{1}{4} e^{4x} \;-\; \int \frac{1}{4} e^{4x}\, dx$

$I \;=\; x\, \frac{1}{4} e^{4x} \;-\; \frac{1}{4} \cdot \frac{1}{4} \int e^{4x}\, 4\, dx$

$I \;=\; \frac{1}{4} x\, e^{4x} \;-\; \frac{1}{16} e^{4x} \;+\; C$

Integration By Parts (IBP) -- Ex. 12

Evaluate the Integral: $\int x e^{-x}\, dx$

Use: L I $\underline{A}$ T E to find u The rest is dv	$u = x$ $du = dx$	$dv = e^{-x}\, dx$ $\int dv = \int e^{-x}\, dx$ $\int dv = (-1)\int e^{-x}(-1)dx$ $v = -e^{-x}$

$I = uv - \int v\, du$

$I = x(-e^{-x}) - \int(-e^{-x})\, dx$

$I = -x e^{-x} + \int e^{-x}\, dx$

$I = -x e^{-x} + (-1)\int e^{-x}(-1)\, dx$

$I = -x e^{-x} - e^{-x}$

$I = -e^{-x}(x + 1) + C$

Integration By Parts (IBP) -- Ex. 13

Evaluate the Integral: $\int x \cos 5x \, dx$

Use: L I **A** T E to find **u**	$u = x$	$dv = \cos 5x \, dx$
	$du = dx$	$\int dv = \frac{1}{5} \int \cos 5x \, (5) \, dx$
The rest is dv		$v = \frac{1}{5} \sin 5x$

$I = uv - \int v \, du$

$I = x \left(\frac{1}{5} \sin 5x \right) - \int \left(\frac{1}{5} \sin 5x \right) dx$

$I = \frac{1}{5} x \sin 5x + \frac{1}{5} \cdot \frac{1}{5} \int \sin 5x \, (5) \, dx$

$I = \frac{1}{5} x \sin 5x + \frac{1}{25} (- \cos 5x)$

$I = \frac{1}{5} x \sin 5x - \frac{1}{25} \cos 5x + C$

Integration By Parts (IBP) -- Ex. 14

Evaluate the Integral: $\int x^2 \cos x \; dx$

Use: L I **A** T E to find **u**	$u = x^2$ $du = 2x \; dx$	$dv = \cos x \; dx$ $\int dv = \int \cos x \; dx$ $v = \sin x$

$I \; = \; uv - \int v \; du$

$I \; = \; x^2 \sin x \; - \; \int \sin x \cdot 2x \; dx$

Use: L I **A** T E to find **u**	$u = 2x$ $du = 2 \; dx$	$dv = \sin x \; dx$ $v = - \cos x$

$I \; = \; x^2 \sin x \; - \; [\, 2x \, (- \cos x) - \int (- \cos x) \, 2 \; dx \,]$

$I \; = \; x^2 \sin x \; + \; 2x \cos x \; - \; 2 \int \cos x \, dx$

$I \; = \; x^2 \sin x \; + \; 2x \cos x \; - \; 2 \sin x$

$I \; = \; \sin x \, (x^2 - 2) \; + \; 2x \cos x \; + \; C$

Integration By Parts (IBP) -- Ex. 15

Evaluate the Integral: $\int x \ln x \, dx$

Use: <u>L</u> I A T E to find $\boldsymbol{u}$ The rest is $\boldsymbol{dv}$	$u = \ln x$ $du = \frac{1}{x} \, dx$	$dv = x \, dx$ $\int dv = \int x \, dx$ $v = \frac{1}{2}x^2$

$I = uv - \int v \, du$

$I = \ln x \left(\frac{1}{2}x^2\right) - \int \left(\frac{1}{2}x^2\right)\frac{1}{x} \, dx$

$I = \frac{1}{2} x^2 \ln x - \frac{1}{2} \int x \, dx$

$I = \frac{1}{2} x^2 \ln x - \frac{1}{2} \cdot \frac{1}{2} x^2$

$I = \frac{1}{2} x^2 \ln x - \frac{1}{4} \cdot x^2$

$I = x^2 \left(\frac{1}{2} \ln x - \frac{1}{4}\right)$

$I = x^2 \left(\ln \sqrt{x} - \frac{1}{4}\right) + C$

IBP – Tabular Method

IBP Tabular Method -- Process (1/2)
$$\int u\, dv \;=\; uv - \int v\, du$$
The Integration By Parts (IBP) Tabular Method is an easier and more efficient way of using IBP. The process is outlined below.

1.	Use the acronym "LIATE" to identify the u part. The rest is the dv part.
2.	Create a table, with u at the top of the 1st column. For each row in the 1st column, keep taking the derivative until one of the following happens: • The derivative $= 0$. • You can't take the derivative. • u was a log function, so just stop after taking the first derivative. • The derivative $= n \cdot u$, a multiple of the original u part.

Continued ...

IBP Tabular Method -- Process (2/2)	
$$\int u\,dv \;=\; uv - \int v\,du$$	

Continued ...

3.	Put the dv part at the top of the 2nd column. For each row in the 2nd column, keep taking the integral.
4.	Between the two columns, put alternating $+/-$ signs. Start with $+$
5.	Draw short, diagonal arrows connecting rows in 1st column to one lower row in the 2nd column.
6.	In last row draw a horizontal arrow.
7.	Write the answer by following the arrows. Multiply diagonal arrows and add or subtract, based on the sign. For horizontal arrows, take the integral.

IBP Tabular Method -- Ex. 1

Evaluate: $I = \int (2x)\cos(9x)\,dx$

Use L I <u>A</u> T E to find u. Setup table.

Du		$\int dv$
$2x$	$+$	$\cos(9x)\,dx$
2	$-$	$\dfrac{1}{9}\sin(9x)$
0	$+$	$\dfrac{-1}{81}\cos(9x)$
Take derivative of this side until you can't or it's $= 0$		Keep taking integral of this side.

$$I = 2x\left(\tfrac{1}{9}\right)\sin 9x \; + \; \tfrac{2}{81}\cos 9x \; + \; \int 0\,dx$$

Note: Last horizontal arrow ➜ Integral

IBP Tabular Method -- Ex. 2

Evaluate: $I = \int t^2 e^t \, dt$

Du	$\int dv$
t^2 $\quad$ $+$	$e^t dt$
$2t$ $\quad$ $-$	e^t
2 $\quad$ $+$	e^t
$\quad$ $-$	e^t
Take derivative of this side until you can't or it's $= 0$	Keep taking integral of this side.

$$I = t^2 e^t - 2te^t + 2e^t - \int e^t \cdot 0 \, dx$$
$$I = t^2 e^t - 2te^t + 2e^t + C$$

Note: Last horizontal arrow ➜ Integral

IBP Tabular Method -- Ex. 3
Evaluate: $I = \int x^5 \ln x \; dx$

Du	$\int dv$
$\ln x$ $\quad +$	$x^5 dx$
$\dfrac{1}{x}$ $\quad -$	$\dfrac{x^6}{6}$
Stop here with logs.	

$$I \;=\; \ln x \left(\tfrac{1}{6}\right) x^6 \;-\; \int \left(\tfrac{1}{6}\right) x^5 \, dx$$

$$I \;=\; \tfrac{1}{6}\left[\, x^6 \ln x \;-\; \int x^5 \, dx \,\right]$$

$$I \;=\; \tfrac{1}{6}\left[\, x^6 \ln x \;-\; \frac{x^6}{6} \,\right]$$

$$I \;=\; \frac{x^6 \ln x}{6} \;-\; \frac{x^6}{36} \;+\; C$$

Note: Last horizontal arrow → Integral

IBP Tabular Method -- Ex. 4
Evaluate: $I = \int e^{2x} \cos 3x \; dx$

Du		$\int dv$
$\cos 3x$	$+$	$e^{2x} dx$
$- 3 \sin 3x$	$-$	$\left(\frac{1}{2}\right) e^{2x}$
$- 9 \cos 3x$	$+$	$\left(\frac{1}{4}\right) e^{2x}$
Stop with $n \cdot u$		

$$I = \frac{(\cos 3x)\, e^{2x}}{2} + \frac{3(\sin 3x)\, e^{2x}}{4}$$
$$+ \int \left(\frac{-9}{4}\right)(\cos 3x) e^{2x} dx$$
$$I = \frac{(\cos 3x)\, e^{2x}}{2} + \frac{3(\sin 3x)\, e^{2x}}{4} - \left(\frac{9}{4}\right) I$$
$$\left(\frac{13}{4}\right) I = \frac{(\cos 3x)\, e^{2x}}{2} + \frac{3(\sin 3x)\, e^{2x}}{4}$$
$$I = \left(\frac{4}{13}\right)\left[\frac{(\cos 3x)\, e^{2x}}{2} + \frac{3(\sin 3x)\, e^{2x}}{4}\right]$$

IBP Tabular Method -- Ex. 5a

Evaluate: $I = \int x^4 e^{2x} \, dx$

Du	$\int dv$
x^4 $\quad+$	$e^{2x} dx$
$4x^3$ $\quad-$	$\left(\dfrac{1}{2}\right) e^{2x}$
$12x^2$ $\quad+$	$\left(\dfrac{1}{4}\right) e^{2x}$
$24x$ $\quad-$	$\left(\dfrac{1}{8}\right) e^{2x}$
24 $\quad+$	$\left(\dfrac{1}{16}\right) e^{2x}$
0 $\quad-$	$\left(\dfrac{1}{32}\right) e^{2x}$
Take derivative of this side until it's $= 0$	Keep taking integral of this side.

$I = $ Continued ...

IBP Tabular Method -- Ex. 5b

Evaluate: $I = \int x^4 e^{2x} \, dx$

Continued from previous page ...

$$I = \frac{x^4 e^{2x}}{2} - \frac{4x^3 e^{2x}}{4} + \frac{12x^2 e^{2x}}{8} \; \cdots$$

$$- \frac{24x \, e^{2x}}{16} + \frac{24 \, e^{2x}}{32}$$

$$I = e^{2x} \left[\frac{x^4}{2} - \frac{4x^3}{4} + \frac{12x^2}{8} - \frac{24x}{16} + \frac{24}{32} \right]$$

$$I = e^{2x} \left[\frac{x^4}{2} - x^3 + \frac{3x^2}{2} - \frac{3x}{2} + \frac{3}{4} \right] + C$$

IBP Tabular Method -- Ex. 6
Evaluate: $I = \int \ln x \; dx$

Du		$\int dv$
$\ln x$	$+$	$1 \; dx$
$\dfrac{1}{x}$	$-$	x
Stop here with logs.		

$I = x \ln x - \int \left(\dfrac{x}{x}\right) dx$

$I = x \ln x - \int (1) \, dx$

$I = x \ln x - x$

$I = x \ln x - x + C$

Note: Last horizontal arrow $\rightarrow$ Integral

IBP Tabular Method -- Ex. 7
Evaluate: $\quad I = \int x \sin x \; dx$

Du	$\int dv$
$x \qquad +$	$\sin x \; dx$
$1 \qquad -$	$- \cos x$
$0 \qquad +$	$- \sin x$
Take derivative of this side until you can't or it's $= 0$	

$$I \;=\; x(-\cos x) \;+\; \sin x \;-\; \int (0)(\sin x)\;dx$$

$$I \;=\; x(-\cos x) \;+\; \sin x \;-\; 0$$

$$I \;=\; -x \cos x \;+\; \sin x \;+\; C$$

IBP Tabular Method -- Ex. 8

Evaluate the Integral: $\int e^{-3x} \sin 5x\, dx$

Du		$\int dv$
$\sin 5x$	$+$	e^{-3x}
$5\cos 5x$	$-$	$\left(-\frac{1}{3}\right) e^{-3x}$
$-25\sin 5x$	$+$	$\left(-\frac{1}{3}\right)^2 e^{-3x}$
Multiple of "u" $\rightarrow$ STOP		

$$I = \sin 5x \left(-\frac{1}{3}\right) e^{-3x} - 5\cos 5x \left(\frac{1}{9}\right) e^{-3x}$$
$$+ \int\left(\frac{1}{9}\right) e^{-3x}\, 25 \sin 5x\, dx$$

$$I = -\frac{1}{3}\sin 5x\, e^{-3x} - \frac{5}{9}\cos 5x\, e^{-3x} - \frac{25}{9}\int e^{-3x}\sin 5x\, dx$$

$$I = -\frac{1}{3}\sin 5x\, e^{-3x} - \frac{5}{9}\cos 5x\, e^{-3x} - \frac{25}{9}I$$

$$\left(\frac{34}{9}\right)I = -\frac{1}{3}\sin 5x\, e^{-3x} - \frac{5}{9}\cos 5x\, e^{-3x}$$

$$I = \frac{9}{34}\left(-\frac{1}{3}\sin 5x\, e^{-3x} - \frac{5}{9}\cos 5x\, e^{-3x}\right)$$

$$I = -\frac{3}{34}\left(\sin 5x\, e^{-3x} + \frac{5}{3}\cos 5x\, e^{-3x}\right) + C$$

Differentials – dx, dt, dy

Differential of the Integral: $dx, dy, or\ dt$

$$\text{Integral} = \int_{z=a}^{z=b} f(z)\ dz$$

- dz is called the **differential** of the variable z .
  It indicates the variable of integration is z .
- $f(z)$ is the **integrand**.
- Points a and b are the **bounds** of integration.

Integral	Differential	Example
$\int_{x=a}^{x=b} f(x)\ dx$	dx	
$\int_{t=a}^{t=b} f(t)\ dt$	dt	
$\int_{y=a}^{y=b} f(y)\ dy$	dy	

Why dy ???

Usually, we integrate with respect to x along the x-axis, on the xy plane.

These integrals look like this: $\int_{x=a}^{x=b} f(x) \ dx$

Sometimes, it is necessary to integrate with respect to y along the y-axis, on the xy plane.

These integrals look like this: $\int_{y=a}^{y=b} f(y) \ dy$

Integral $= \int_{x=a}^{x=b} f(x) \ dx$ -- **Example 1**

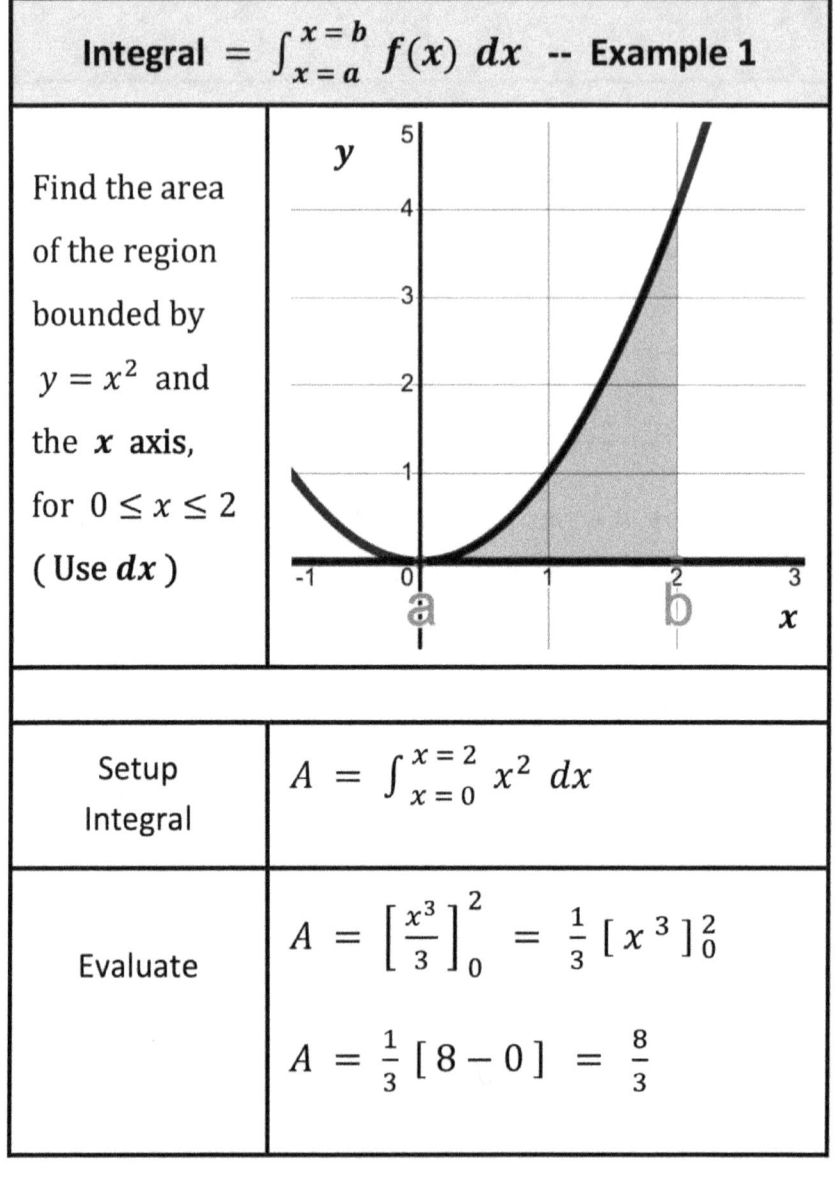

Find the area
of the region
bounded by
$y = x^2$ and
the x axis,
for $0 \leq x \leq 2$
(Use dx)

Setup Integral	$A = \int_{x=0}^{x=2} x^2 \ dx$

Evaluate	$A = \left[\dfrac{x^3}{3} \right]_0^2 = \dfrac{1}{3} \left[x^3 \right]_0^2$
	$A = \dfrac{1}{3} \left[8 - 0 \right] = \dfrac{8}{3}$

Integral $= \int_{y=a}^{y=b} f(y) \, dy$ -- Example 2	
Find the area of the region bounded by $y = x^2$ and the y axis, for $0 \le x \le 2$ (Use dy)	

Rewrite function	$y = x^2 \quad \rightarrow \quad x = \pm\sqrt{y}$ Boundaries $\quad \rightarrow \quad 0 \le y \le 4$
Setup Integral	$A = \int_{y=0}^{y=4} y^{\frac{1}{2}} \, dy$
Evaluate	$A = \left[\left(\frac{2}{3}\right) y^{\frac{3}{2}} \right]_0^4 = \frac{2}{3} \left[y^{\frac{3}{2}} \right]_0^4$ $A = \frac{2}{3} \, [\, 8 - 0 \,] = \frac{16}{3}$

Integral $= \int_{y=a}^{y=b} f(y) \; dy$ -- **Example 3**

Find the area of the region bounded by $y = x^2$ and the line $x = 2$ for $0 \le x \le 2$

Integrate along the y-axis. (Use dy)

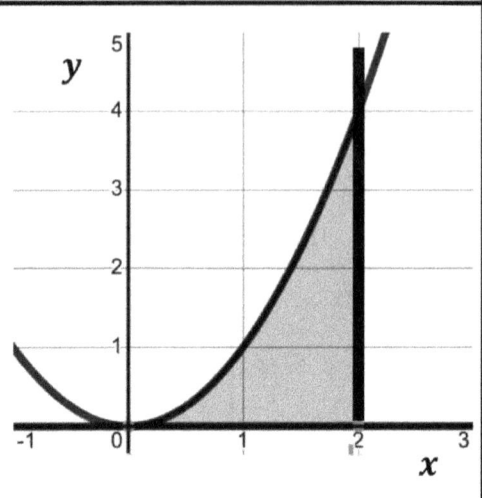

Rewrite function	$y = x^2 \quad \rightarrow \quad x = \pm\sqrt{y}$
	Boundaries $\rightarrow 0 \le y \le 4$
Setup Integral	$A = \int_{y=0}^{y=4} \left(y - y^{\frac{1}{2}} \right) dy$

$$A = \left[\left(\tfrac{1}{2}\right) y^2 - \left(\tfrac{2}{3}\right) y^{\frac{3}{2}} \right]_0^4$$

$$A = \left[(8) - \left(\tfrac{16}{3}\right) \right] = \tfrac{8}{3}$$

Same answer as Example 1

Applications of Integrals

Areas Between Curves

Area Between 2 Curves

$$A = \int_a^b [f(x) - g(x)]\, dx$$

Is the area of the region bounded by

Curves: $y = f(x)$ and $y = g(x)$

Lines: $x = a$ and $x = b$

Where: $f(x) \geq g(x)$ for all x in $[a, b]$

On the interval $[\,a, b\,]$	Between Intersection Pts.

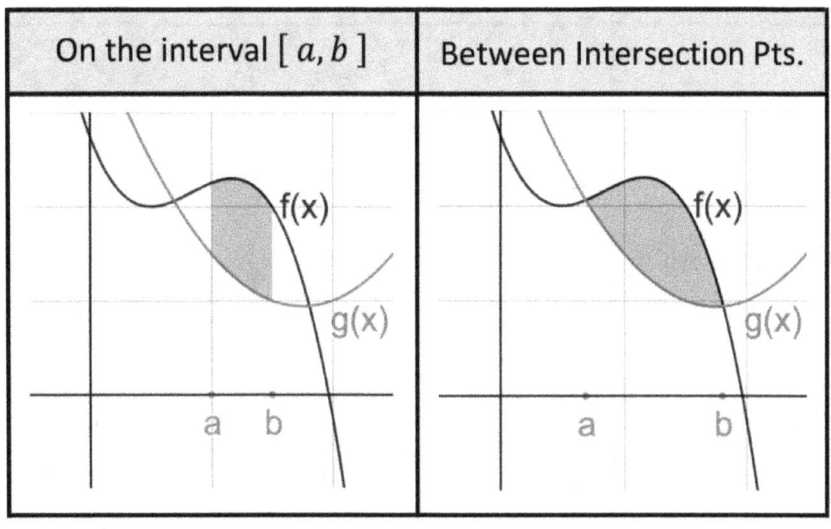

Area Between 2 Curves -- Ex. 1

Find the area of the region bounded by:

$y = e^x$ (above) and

$y = x$ (below)

And bounded on sides by:

$x = 0$ and $x = 1$.

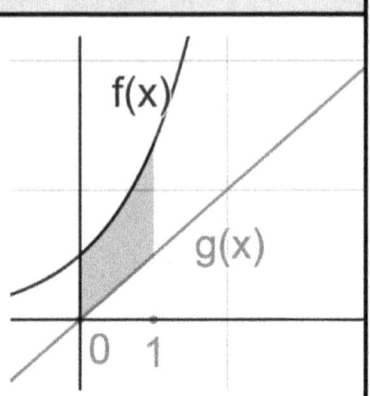

Use Area equation.	$A = \int_a^b [f(x) - g(x)]\ dx$
Find the area.	$A = \int_0^1 [e^x - x]\ dx$ $A = \left[e^x - \dfrac{x^2}{2} \right]_0^1\ dx$ $A = \left[\left(e^1 - \dfrac{1^2}{2}\right) - \left(e^0 - \dfrac{0^2}{2}\right) \right]$ $A = \left[\left(e - \dfrac{1}{2}\right) - (1 - 0) \right]$ $A = \left[e - \dfrac{1}{2} - 1 \right]$ $A = \left[e - \dfrac{3}{4} \right]$

Area Between 2 Curves -- Ex. 2

Find the area of the region bounded by 2 curves:

$f(x) = 2x - x^2$ and

$g(x) = x^2$

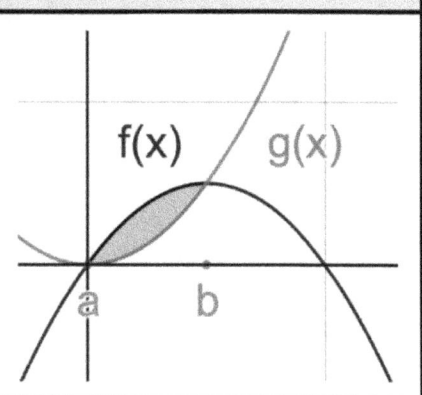

Find the intersect points.	$2x - x^2 = x^2$ $2x - 2x^2 = 0$ $2x(1 - x) = 0 \quad \rightarrow \quad x = 0, 1$
Find the area between intersect points.	$A = \int_0^1 [\,(2x - x^2) - (x^2)\,]\;dx$ $A = \int_0^1 [\,2x - 2x^2\,]\;dx$ $A = 2\int_0^1 [\,x - x^2\,]\;dx$ $A = 2\left[\dfrac{x^2}{2} - \dfrac{x^3}{3}\right]_0^1 \;dx$ $A = 2\left[\left(\dfrac{1}{2} - \dfrac{1}{3}\right) - 0\right] = \dfrac{1}{3}$

Area Between 2 Curves -- Ex. 3 $\int dx$

Find the area of the region bounded by 2 curves:

$$f(x) = (x - 4)^2 + 2$$

$$g(x) = 11$$

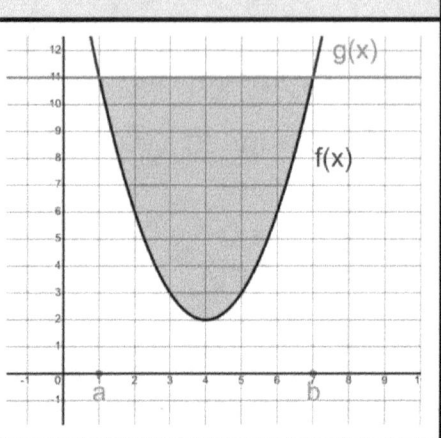

Find the intersect points.	$(x - 4)^2 + 2 = 11$ $x - 4 = \pm\sqrt{9} = \pm 3 = 1, 7$
Find the area, using intersect points as boundaries	$A = \int_1^7 [11 - ((x - 4)^2 + 2)]\, dx$ $= \int_1^7 [9 - (x^2 - 8x + 16)]\, dx$ $= \int_1^7 [-x^2 + 8x - 7]\, dx$ $= \left[-\dfrac{x^3}{3} + \dfrac{8x^2}{2} - 7x \right]_1^7$ $= \left[\dfrac{98}{3} - \left(-\dfrac{10}{3}\right) \right] = \dfrac{108}{3} = 36$

Area Between 2 Curves -- Ex. 4	$\int dy$

Find the area of the region bounded by 2 curves:

$f(y) = (y - 4)^2 + 2$

$g(y) = 11$

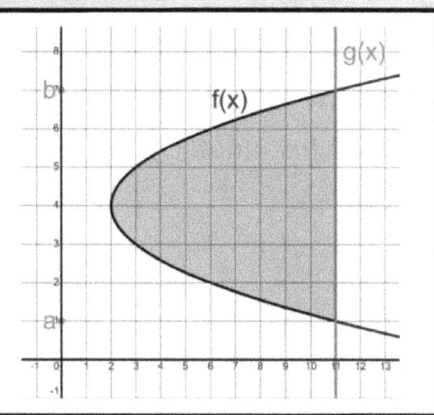

Find the intersect points.	$(y - 4)^2 + 2 = 11$ $y - 4 = \pm\sqrt{9} = \pm 3 = 1, 7$
Find the area, using intersect points as bound- aries.	$A = = \int_1^7 [\, 11 - ((y - 4)^2 + 2)\,]\, dy$ $= \int_1^7 [\, 9 - (y^2 - 8y + 16)\,]\, dy$ $= \int_1^7 [\, -y^2 + 8y - 7\,]\, dy$ $= \left[-\dfrac{y^3}{3} + \dfrac{8y^2}{2} - 7y \right]_1^7$ $= \left[\dfrac{98}{3} - \left(-\dfrac{10}{3} \right) \right] = \dfrac{108}{3} = 36$

Area Between 2 Curves -- Ex. 5	$\int dx$

Find the area of the region bounded by 2 curves:

$y_1 = \sqrt{x}$ and

$y_2 = x$

Find the intersect points.	$x = \sqrt{x}$ $x^2 = x$ $x(x-1) = 0$ $\rightarrow$ $x = 0, 1$
Find the area, using intersect points as boundaries	$A = \ = \int_{x=0}^{x=1} \left(x^{\frac{1}{2}} - x \right) dx$ $= \left[\left(\frac{2}{3}\right) x^{\frac{3}{2}} - \frac{x^2}{2} \right]_0^1$ $= \left[\frac{2}{3} - \frac{1}{2} \right] = \left[\frac{4}{6} - \frac{3}{6} \right] = \frac{1}{6}$

Area Between 2 Curves -- Ex. 6 $\int dy$

Find the area of the region bounded by 2 curves:

$$y_1 = \sqrt{x} \quad \& \quad y_2 = x$$

Integrate with respect to the y-axis.

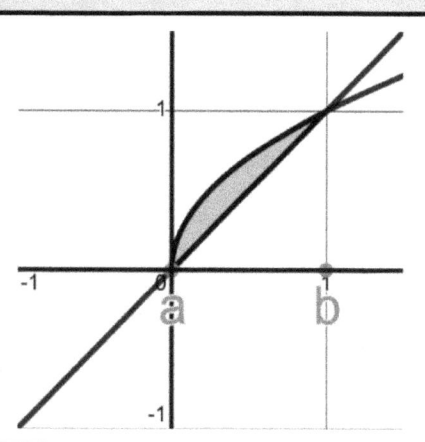

Convert equations	We want: $x = f(y)$ $$x_1 = y^2 \qquad \text{and} \qquad x_2 = y$$
Find the intersect points.	$y^2 = y$ $$y(y - 1) = 0 \quad \rightarrow \quad y = 0, 1$$
Find the area, using intersect points as boundaries	$A = \int_{y=0}^{y=1}(y - y^2)\, dy$ $y \geq y^2$ in the interval $$A = \left[\frac{y^2}{2} - \frac{y^3}{3} \right]_0^1$$ $$A = \left[\frac{1}{2} - \frac{1}{3} \right] = \left[\frac{3}{6} - \frac{2}{6} \right] = \frac{1}{6}$$

Area Between 2 Curves -- Ex. 7a

Find the area (A_1)
of the region
bounded by
$y = x^2$
and the x axis,
for $0 \le x \le 1$

Setup Integral	$A_1 = \int_{x=0}^{x=1} x^2\, dx$
Evaluate	$A_1 = \left[\frac{x^3}{3}\right]_0^1 = \frac{1}{3}[x^3]_0^1$
	$A_1 = \frac{1}{3}[1-0] = \frac{1}{3}$

Continued ...

Area Between 2 Curves -- Ex. 7b

Find the area (A_2)
of the region
bounded by

$y = x^2$

and the y axis,

for $0 \le y \le 1$

Rewrite function	$y = x^2 \qquad \rightarrow \qquad x = \pm\sqrt{y}$
Setup Integral	$A_2 = \int_{y=0}^{y=1} y^{\frac{1}{2}} \, dy$
Evaluate	$A_2 = \left[\left(\frac{2}{3}\right) y^{\frac{3}{2}} \right]_0^1 = \frac{2}{3} \left[y^{\frac{3}{2}} \right]_0^1$ $A_2 = \frac{2}{3} [1 - 0] = \frac{2}{3}$
Notice:	$A_1 + A_2 = \frac{1}{3} + \frac{2}{3} = 1$

Volumes by Disks

Volume -- Disks

Rotate a curve about the x-axis to create a cylindric solid. Cut into circular disks with a thickness of dx.

$$V = \sum (Area)\ (Thickness)$$

$$V = \int_a^b A(x)\ dx$$

Where $A(x)$ is the cross sectional area of the solid and <u>perpendicular</u> to the x-axis.

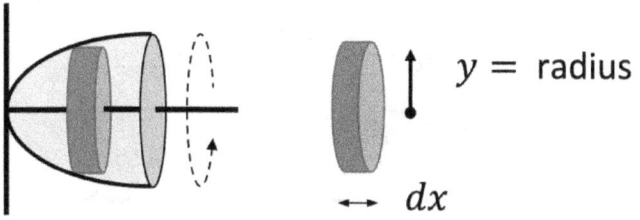

$y = $ radius

dx

Volume -- Disks

$$V = \int_a^b A(x)\ dx$$

Where $A(x)$ is the cross sectional area of the solid and <u>perpendicular</u> to the x-axis.

In other words ...

$$V = \text{Volume of a cylinder}$$

$$V = (\text{Base Area}) * (\text{Thickness})$$

$$V = \int_a^b (Base\ Area)\ dx$$

$$V = \int_a^b (\pi\, r^2)\ dx$$

$$V = \int_a^b (\pi\, y^2)\ dx$$

$$V = \int_a^b A(x)\ dx$$

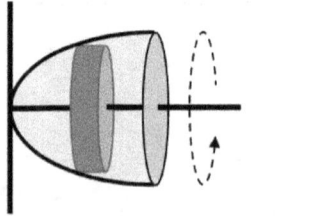

$y = \text{radius}$

dx

Volume (Disks) -- Ex. 1

Find the volume of a solid created by rotating $f(x)$ about the x-axis for:

$$y = f(x) = 3$$

on $[2, 7]$

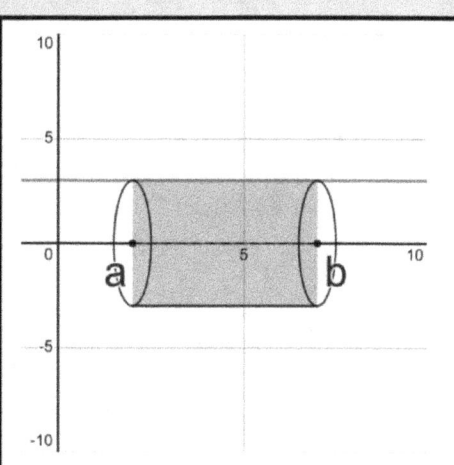

Setup the integral.	$V = \int_a^b (Base\ Area)\, dx$
	$V = \int_2^7 (\pi r^2)\, dx$
	$V = \int_2^7 (\pi y^2)\, dx$
	$V = \int_2^7 (\pi \cdot 3^2)\, dx$
Evaluate the integral.	$V = 9\pi \int_2^7 (1)\, dx$
	$V = 9\pi\, [\, x\,]_2^7$
	$V = 9\pi\, [7 - 2] = 45\pi \approx 141.4$

Volume (Disks) -- Ex. 2

Find the volume of

a solid created by

rotating $f(x)$ about

the x-axis for:

$$y = f(x) = x$$

on $[0, 5]$

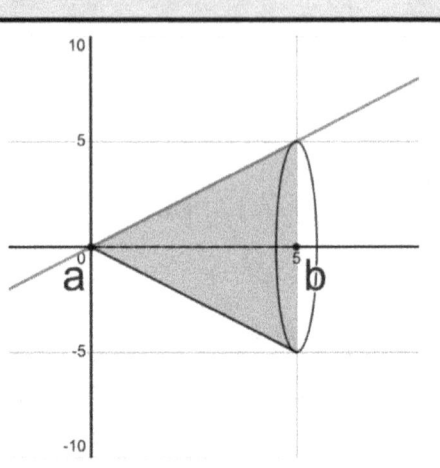

Setup the integral.	$V = \int_a^b (Base\ Area)\ dx$
	$V = \int_0^5 (\pi r^2)\ dx$
	$V = \int_0^5 (\pi y^2)\ dx$
	$V = \int_0^5 (\pi \cdot x^2)\ dx$
Evaluate the integral.	$V = \pi \int_0^5 (x^2)\ dx$
	$V = \pi \left[\dfrac{x^3}{3}\right]_0^5$
	$V = \dfrac{\pi}{3}[5^3 - 0] = \dfrac{125\pi}{3} \approx 130.9$

Volume (Disks) -- Ex. 3

Find the volume of a solid created by rotating $f(x)$ about the x-axis for:

$$y = f(x) = x$$

on $[2, 5]$

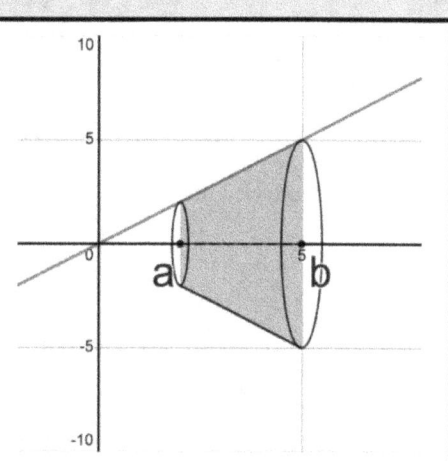

Setup the integral.	$V = \int_a^b (Base\ Area)\,dx$
	$V = \int_2^5 (\pi r^2)\,dx$
	$V = \int_2^5 (\pi y^2)\,dx$
	$V = \int_2^5 (\pi \cdot x^2)\,dx$
Evaluate the integral.	$V = \pi \int_2^5 (x^2)\,dx$
	$V = \pi \left[\dfrac{x^3}{3}\right]_2^5$
	$V = \dfrac{\pi}{3}[5^3 - 2^3] = \dfrac{117\pi}{3} \approx 122.5$

Volume (Disks) -- Ex. 4	
Find the volume of a solid created by rotating $f(x)$ about the x-axis for: $$y = f(x) = \frac{x^2}{2}$$ on $[\,2, 5\,]$	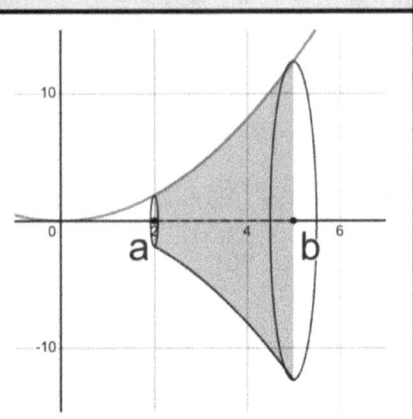

Setup the integral.	$V = \int_a^b (Base\ Area)\,dx$ $V = \int_2^5 (\pi r^2)\,dx$ $V = \int_2^5 (\pi y^2)\,dx$ $V = \int_2^5 \left(\pi \cdot \left(\frac{x^2}{2}\right)^2\right)\,dx$
Evaluate the integral.	$V = \frac{\pi}{4} \int_2^5 (x^4)\,dx$ $V = \frac{\pi}{4}\left[\frac{x^5}{5}\right]_2^5$ $V = \frac{\pi}{20}[\,5^5 - 2^5\,] = \dfrac{3093\,\pi}{20}$

Volumes by Cylindrical Shells

Volume – Cylindrical Shells

Rotate the region, bound by 2 curves, about the y-axis to create a cylindrical solid. Cut into cylindrical shells with a thickness of dx.

$$V = \sum \text{Volume of a Cylinder}$$

$$V = \sum (\text{Circumference})*(\text{Height})*(\text{Thickness})$$

$$V = \int_a^b (2\pi r) \cdot (y_1 - y_2)\ dx$$

$$V = \int_a^b (2\pi x) \cdot [f_1(x) - f_2(x)]\ dx$$

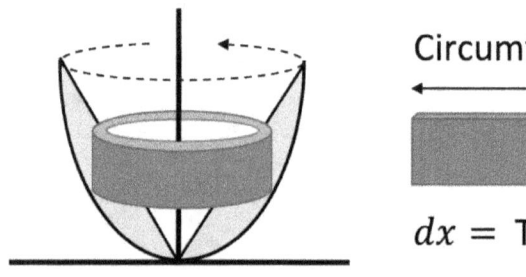

Circumference

h

$dx = \text{Thickness}$

Volume – Cylindrical Shells

$$V = \int_a^b 2\pi x \, f(x) \, dx$$

The volume of the solid is obtained by rotating a region, bounded by 2 curves, about the y-axis.

In other words ...

$V = \sum$ Volume of a cylinder

$V = \sum$ (Circum.)*(Height)*(Thickness)

$V = \int_a^b (2\pi r) \cdot (y_1 - y_2) \, dx$

$V = \int_a^b (2\pi x) \cdot [\, f_1(x) - f_2(x)] \, dx$

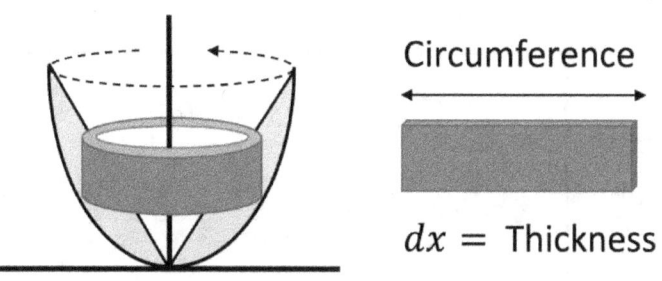

Circumference

h

dx = Thickness

Volume (Cylindrical Shells) -- Ex. 1

Find the volume of a solid created by rotating, about the y-axis, the region bounded by:

$y_1 = x$ and $y_2 = x^2$

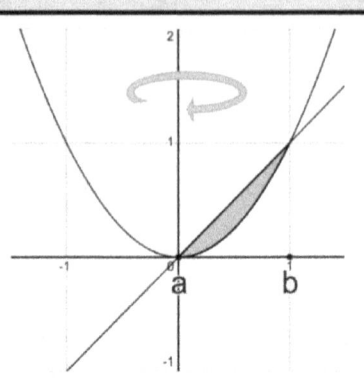

Find Intersect points.	$x = x^2$ $x - x^2 = 0$ $x(1 - x) = 0 \qquad \rightarrow \qquad x = 0, 1$
Setup integral.	$V = \int_a^b (2\pi r)\,(y_1 - y_2)\,dx$ $V = \int_0^1 (2\pi x)\,(x - x^2)\,dx$
Evaluate integral.	$V = 2\pi \int_0^1 (x^2 - x^3)\,dx$ $V = 2\pi \left[\dfrac{x^3}{3} - \dfrac{x^4}{4} \right]_0^1 = 2\pi \left[\dfrac{1}{3} - \dfrac{1}{4} \right]$ $V = 2\pi \left[\dfrac{4}{12} - \dfrac{3}{12} \right] = \dfrac{2\pi}{12} = \dfrac{\pi}{6}$

Volume (Cylindrical Shells) -- Ex. 2

Find the volume of a solid created by rotating, about the y-axis, the region bounded by:

$y_1 = 3$ and $y_2 = \sqrt{x}$

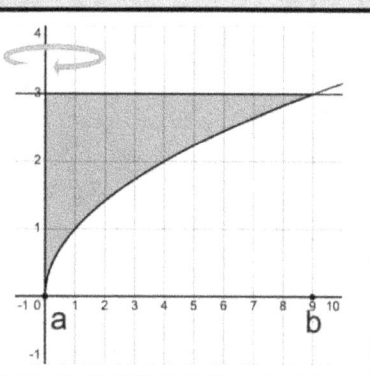

Find Intersect pts.	$3 = \sqrt{x}$ $\rightarrow$ $9 = x$
Setup integral.	$V = \int_a^b (2\pi r)\,(y_1 - y_2)\,dx$ $V = \int_0^9 (2\pi x)\left(3 - \sqrt{x}\right) dx$
Evaluate integral.	$V = 2\pi \int_0^9 \left(3x - x^{\frac{3}{2}}\right) dx$ $V = 2\pi \left[\left(\frac{3}{2}\right) x^2 - \left(\frac{2}{5}\right) x^{\frac{5}{2}} \right]_0^9$ $V = 2\pi \left[\left(\frac{3}{2}\right) 9^2 - \left(\frac{2}{5}\right) 9^{\frac{5}{2}} \right]$ $V = 2\pi \left[\dfrac{243}{10} \right] = \dfrac{243\,\pi}{5}$

Work

Work Equation
$$W \ = \ \int_a^b F(x) \ dx$$ Where $F(x)$ is the applied force.
In other words ... Work is the force applied over a distance, from a to b.

Work Terms

Terms	SI Units	English Units
m = mass	kg.	lb.
a = accel.	$\left[\dfrac{m}{s^2}\right]$	$\left[\dfrac{ft.}{s^2}\right]$
d = Dist.	m = meter	ft. = feet
F = Force $\\ F = ma$	N = Newton $\left[\dfrac{kg \cdot m}{s^2}\right]$	Pound $\approx 4.45\ N$
W = Work $\\ W = Fd$	J = Joule $[\,N\,m\,]$	Foot-pound $\approx 1.36\ J$
t = time	sec.	sec.
g = gravity	$9.8\left[\dfrac{m}{s^2}\right]$	$32\left[\dfrac{ft.}{s^2}\right]$

	Work -- Ex. 1
	How much work is done, moving a particle from $x = 1$ ft. to $x = 4$ ft. if the force is: $f(x) = x^2 + 2x$ lbs.

Setup the integral.	$W = \int_a^b f(x)\, dx$ $W = \int_1^4 (x^2 + 2x)\, dx$
Calculate Work	$W = \left[\dfrac{x^3}{3} + 2\dfrac{x^2}{2}\right]_1^4$ $W = \left[\dfrac{x^3}{3} + x^2\right]_1^4$ $W = \dfrac{1}{3}[\, x^3 + 3x^2\,]_1^4$ $W = \dfrac{1}{3}[(4^3 + 3\cdot 4^2) - (1 + 3)]$ $W = \dfrac{1}{3}[\,(112) - (4)]$ $W = \dfrac{1}{3}[108] \;=\; 36$ ft-lb

Work -- Ex. 2a

A 100 ft. cable weighs 300 lbs. and is hanging from the top of a tall building. How much work is required to lift the cable to the top of the building?

Make a sketch.	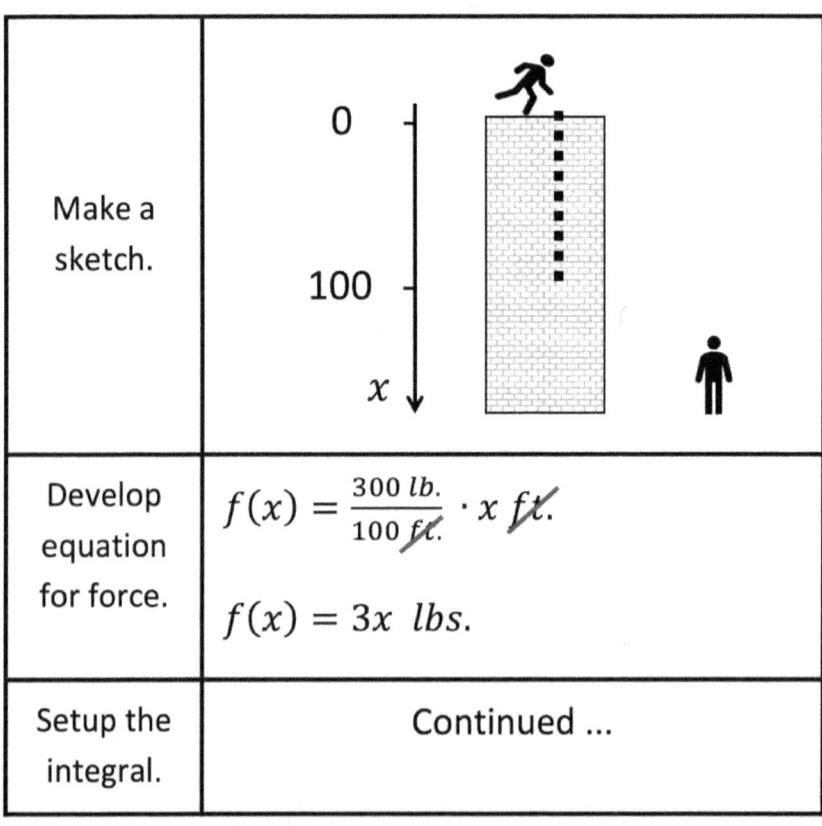
Develop equation for force.	$f(x) = \dfrac{300 \, lb.}{100 \, ft.} \cdot x \, ft.$ $f(x) = 3x \; lbs.$
Setup the integral.	Continued ...

Work -- Ex. 2b	

A 100 ft. cable weighs 300 lbs. and is hanging from the top of a tall building. How much work is required to lift the cable to the top of the building?

Previously found	Force $= f(x) = 3x$ lbs. $a = 0$ and $b = 100$ ft.
Setup the integral.	$W = \int_a^b f(x)\ dx$ $W = \int_0^{100} (3x)\ dx$
Evaluate the integral.	$W = 3\left[\dfrac{x^2}{2}\right]_0^{100}$ $W = \dfrac{3}{2}\left[x^2\right]_0^{100}$ $W = \dfrac{3}{2}\left[(100)^2 - 0\right]$ $W = 15{,}000$ ft-lb

<u>Average Value of a Function</u>

Average Value Equation
$$f_{avg} = \frac{1}{b-a} \int_a^b f(x)\ dx$$ Average value of f on $[\,a, b\,]$
In other words …   Finding the average of an integral on a closed interval is similar to finding the average of a set of numbers. $$Avg. = \frac{a_1 + a_2 + a_3 + \ldots + a_n}{n}$$

Mean Value Theorem (MVT)
For Integrals

If f is continuous on $[\,a, b\,]$

then there is a number c in $[\,a, b\,]$ such that:

$$f(c) \;=\; f_{avg} \;=\; \frac{1}{b-a} \int_a^b f(x)\ dx$$

$$\int_a^b f(x)\ dx \;=\; f(c)\,(b-a)$$

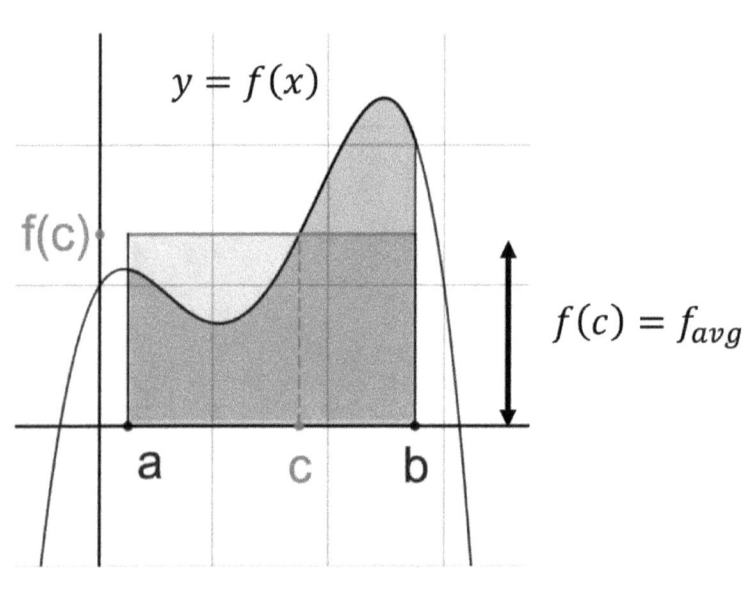

$y = f(x)$

f(c)

$f(c) = f_{avg}$

a   c   b

Average Value -- Ex. 1a
Find the average value of the function $f(x) = 1 + x^2$ on the interval $[-1, 5]$. Also, find c in the interval, where $\quad f_{avg} = f(c)$

$f_{avg} = \dfrac{1}{b-a} \int_a^b f(x)\, dx$ $f_{avg} = \dfrac{1}{5+1} \int_{-1}^{5} (1 + x^2)\, dx$
$f_{avg} = \dfrac{1}{6}\left[x + \dfrac{x^3}{3} \right]_{-1}^{5}$ $f_{avg} = \dfrac{1}{6}\left[\left(5 + \dfrac{125}{3} \right) - \left(-1 - \dfrac{1}{3} \right) \right]$ $f_{avg} = \dfrac{1}{6}\left[\dfrac{140}{3} + \dfrac{4}{3} \right] = \dfrac{1}{6}[48] = 8$ Note: Area Under curve $= 48$

Continued ...

Average Value -- Ex. 1b	
Find the average value of the function $f(x) = 1 + x^2$ on the interval $[-1, 5]$. Also, find c in the interval, where $f_{avg} = f(c)$	

Previously Found	$f_{avg} = f(c) = 8$
Find "c"	$f(x) = 1 + x^2$ $8 = 1 + x^2$ $x = \pm\sqrt{7}$ $c = \sqrt{7} \approx 2.6$ $-\sqrt{7}$ not in interval.

Continued ...

Average Value -- Ex. 1c

Find the average value of the function

$f(x) = 1 + x^2$ on the interval $[-1, 5]$.

Also, find c in the interval, where $f_{avg} = f(c)$

Previously Found	$f_{avg} = f(c) = 8$
	$c = \sqrt{7} \approx 2.6$
	$\int_a^b f(x)\, dx = 48$

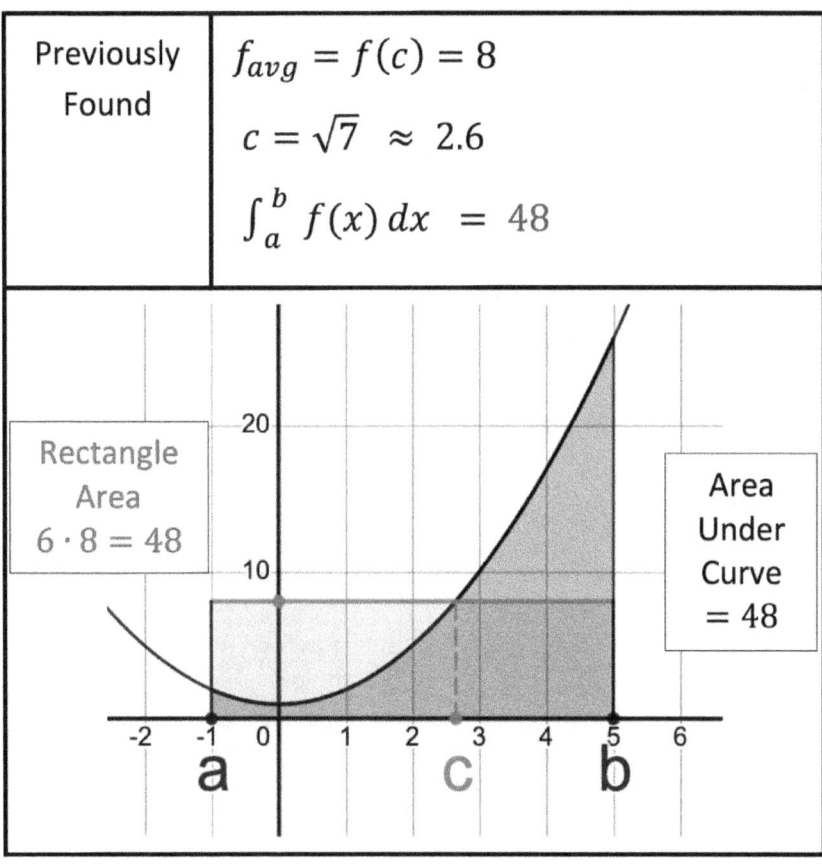

Rectangle Area

$6 \cdot 8 = 48$

Area Under Curve $= 48$

References

References

- Calculus Early Transcendentals,
  Eighth Edition, 2015, James Stewart.

- Calculus 10e, Tenth Edition, 2014,
  Ron Larson, Bruce Edwards.

- Calculus Early Transcendentals Single Variable,
  Ninth Edition, 2009,
  Howard Anton, Irl Bivens, Stephen Davis.

- Differential Equations With Applications: Class
  Notes With Detailed Examples, 2019,
  Jigarkumar Patel and Kathryn Paulk.

References

- Algebra and Trigonometry, Structure and Method, Book 2, Houghton Mifflin, Richard Brown, Mary Dolciani, Robert Sorgenfrey, Robert Kane, 1992.

- Mathematics, Structure and Method, Course 2, Mary Dolciani, Robert Sorgenfrey, John Grahm, McDougal Littell, 2001.

- Precalculus, A Graphing Approach, 2nd Edition, Larson, Hostetler, Edwards, Houghton Mifflin Company, 1997.

- Essentials of College Algebra, Richard Aufmann, Richard Nation, Houghton Mifflin, 2006.

- One-Page Summaries for Algebra, Geometry, and Pre-Calculus, Kathryn Paulk, 2023.

- Complex Numbers and Polar Curves for Pre-Calc and Trig: With Problems and Detailed Solutions, Kathryn Paulk, 2023.

<u>Other Books by Kathryn Paulk</u>

Other Books by Kathryn Paulk

- Algebra 1 Help
- Algebra 2 Help
- Pre-Calculus and Trig Help
- College Algebra Help

- Calculus 1 Review in Bite-Size Pieces
- Calculus 2 Review in Bite-Size Pieces
- Calculus 3 Review in Bite-Size Pieces
- Differential Equations With Applications: Class Notes With Examples

---

- One-Page Summaries for Algebra, Geometry & Pre-Calc.
- Pre-Calculus and Trig Problems & Solutions
- Graphing Functions Using Transformations for Algebra & Pre-Calculus
- Complex Numbers and Polar Curves For Pre-Calc and Trig: With Problems and Detailed Solutions
- Discrete and Continuous Probability Distributions: A Creative Comparison (V2)

- Teach Your Child to SWIM

BIG MATH For Little Kids

Workbooks for Young Children
& Solution Manuals for Parents

- Introduction to Numbers
  (Ages 2 – 5)

- Introduction to Fractions by Sharing Things
  (Ages 3 – 8)

- Introduction to Counting & Fractions by Cooking Breakfast
  (Ages 5 and up)

- Learn About Fractions by Baking Cookies
  (Ages 8 and up)

- Adding Big Numbers, Guessing Numbers and Secret Codes
  (Ages 8 and up)

- Learn to Graph by Riding Bikes on Graph Paper
  (Ages 10 and up)

These books are based on the activities
Kathy did with her son when he was young.

www.ingramcontent.com/pod-product-compliance
Lightning Source LLC
Chambersburg PA
CBHW072144230526
45467CB00040B/23